THE MEDIATED CLIMATE

THE MEDIATED CLIMATE

HOW JOURNALISTS, BIG TECH, AND ACTIVISTS ARE VYING FOR OUR FUTURE

ADRIENNE RUSSELL

Columbia University Press *New York*

Columbia University Press
Publishers Since 1893
New York   Chichester, West Sussex
cup.columbia.edu
Copyright © 2023 Columbia University Press
All rights reserved

Library of Congress Cataloging-in-Publication Data
Names: Russell, Adrienne, author.
Title: The mediated climate : how journalists, big tech, and activists are vying for our future / Adrienne Russell.
Description: New York : Columbia University Press, 2023. | Includes bibliographical references and index.
Identifiers: LCCN 2022058506 (print) | LCCN 2022058507 (ebook) | ISBN 9780231201728 (hardback) | ISBN 9780231201735 (trade paperback) | ISBN 9780231554237 (ebook)
Subjects: LCSH: Climatic changes—Press coverage. | Mass media and the environment. | Misinformation. | Disinformation. | Journalistic ethics.
Classification: LCC QC903 .R865 2023 (print) | LCC QC903 (ebook) | DDC 070.4/4936373874—dc23/eng/20230105
LC record available at https://lccn.loc.gov/2022058506
LC ebook record available at https://lccn.loc.gov/2022058507

Cover design: Julia Kushnirsky
Cover photograph: Elizabeth Weber

For Sofia and Sam

Continue to contaminate your bed, and you will one night suffocate in your own waste. When the buffalo are all slaughtered, the wild horses all tamed, the secret corners of the forest heavy with the scent of men, and the view of the ripe hills blotted by talking wires, where is the thicket? Gone. Where is the eagle? Gone. And what is to say goodbye to the swift and the hunt, the end of living and the beginning of survival.

—Duwamish chief Seattle in a letter to President Franklin Pierce, 1855

CONTENTS

Introduction: Two Crises   1
1 House on Fire   39
2 Noise, Incivility, and Ambivalence   67
3 After Peak Indifference   103
4 Collective Imaginary   133

Acknowledgments   163
Notes   167
Bibliography   211
Index   245

THE MEDIATED CLIMATE

INTRODUCTION

Two Crises

In 2019, Bill McKibben, the American journalist turned founder of the seminal climate activist organization 350.org, told an auditorium crowd at the University of Washington about what he was thinking in 1989 after he published *End of Nature*, his first book on climate crisis. "OK, I did that; now people will understand the danger we face and will start making change to address the problem." McKibben laughed at the memory, and so did the audience. "Yep, that's really what I thought," he added, sparking another round of laughter. The humor, of course, was tied to what seems now an impossible innocence about the nature of today's informational media landscape—a vast and bruising place colonized in great swaths by content of often murky provenance and dubious quality, much of it varied genres of mis- and disinformation that run in all directions, averting public consensus on the facts and working to stymie action on even the gravest matters of public interest.

McKibben's laugh line likely wouldn't have landed with the same kind of punch in 1962, the year the marine biologist, editor, and nature writer Rachel Carson made an enormous impact with *Silent Spring*, her book on the widespread harmful effects of chemical pesticides.[1] *Silent Spring* shined a bright light on a major environmental threat and succeeded in spurring people to make change. It generated support for new environmental policy and sparked renewed awareness about the way corporate interests can run counter to public health and safety, even as it drew now-familiar corporate attacks. The agricultural chemical industry poured vast sums into efforts to discredit the book. Industry operatives dismissed Carson as a "spinster" and called her a communist sympathizer and Soviet propagandist working to reduce food supplies in the West. The mainstream press took jabs, too. A *Time* magazine reviewer criticized Carson for her "emotion-fanning words" and characterized her arguments as "unfair, one-sided, and hysterically overemphatic."[2] But the attacks failed to blunt the power of the book. President John F. Kennedy's Science Advisory Committee launched a special investigation that confirmed Carson's conclusions and began the work of establishing the Environmental Protection Agency. A decade after the book's publication, the U.S. government banned the use of DDT, the powerful pesticide at the center of Carson's investigation.[3]

Carson published *Silent Spring* during the high-modern period of journalism in the United States, between the end of World War II and the 1980s, when, compared to today, the journalism industry labored under less immediate economic pressure, journalists as a group enjoyed a historically high level of professional autonomy, and the government exercised greater regulatory authority over the media industry.[4] The United States also experienced comparatively high levels of ideological consensus at the time, fueling the development of and commitment to the norm of objectivity in professional journalism, which demanded fairness and balance, even if only within the limits of consensus or the "common sense" of a particular ideology and political system.⁵ Those facts mitigated attacks on Carson. Journalists covering the dangers of toxic pesticides and the budding environmental movement closely aligned with scientists and politicians, who were also taking up the cause.

The high-modern period of journalism didn't deliver all positives, of course.⁶ Most notably, its devotion to official sources, hard facts, and editorial balance produced views that served the elite. But Carson's success in taking on the powerful chemical industry had much to do with the fact that the political elite were on her side. Her book pointed fingers at powerful people and demanded significant change, but it also traded on and gained enormous strength from sentiments held by people wielding political power, most notably the increasingly popular view of the importance of conserving nature and protecting public health. Indeed, Carson's work can best be understood as part of a long line of influential twentieth-century investigative journalism that exposed the sources of violence and corruption. Ida B. Wells, for example, reported on the violence against Black Americans during the era of lynching,⁷ and Upton Sinclair uncovered abhorrent labor and sanitary conditions in the meat-packing industry in the 1910s.⁸ These journalists simultaneously challenged powerful elites and lent credibility to the institution of journalism by providing evidence of its dedication to serving as watchdog.

Today, the challenges of ringing alarm bells in a way that people hear and act on them seem different in important ways. We now simultaneously live on a hyperdeveloped planet that is approaching environmental catastrophe and mass species extinction and in a hypermediated age marked by informational war and manufactured mass confusion that afflicts many of the world's longest-running democracies. The climate crisis is an enormous challenge to address because pollution rings the globe in deadly amounts and spews from technologies woven into the everyday fabric of human life. According to the Intergovernmental Panel on Climate Change (IPCC), the crisis is systemic, meaning that it demands changes to the very nature of our existing political, economic, and communication institutions and practices[9]

The crisis is also an enormous challenge because of the way it is integrally linked with racism and inequity on local, national, and global levels. In the United States, for instance, communities of color endure high rates of diabetes from lack of access to healthy food and high rates of asthma from living close to oil refineries and freeways or working with toxic chemicals such as pesticides.[10] Countries in the Global South grapple with the harshest impacts of climate crisis, including extreme weather, flooding, drought, and fire, with little support from wealthier nations. As Robert Bullard, distinguished professor at Texas Southern University who is known as the father of environmental justice, puts it, "The most peculiar aspect of climate change is that the populations that contribute least to the problem of climate change are most likely to feel its impacts. Such disproportionality makes it a serious social justice issue."[11] The climate crisis is a challenge not least because the only way to address it is also to address the racist structures that undergird it, including, of course, the structure of our media environment.[12]

What kind of book, movie, coordinated multigenre, multi-platform campaign would it take today to make a Rachel Carson kind of difference that would rouse a critical mass of various kinds of governments and elites around the world to address the climate crisis?

The Mediated Climate looks at the climate crisis through the lens of communication. It examines key developments in today's information ecosystem that have worked against efforts to turn scientific knowledge about the climate crisis into impactful solutions and explores efforts that we—concerned people of the planet and readers of this book—might take to address the problem. The book focuses on the intersections where the climate and information crises meet to better understand the conditions under which public discourse about the climate crisis has taken shape. By using the climate crisis as a critical lens to examine the forces shaping the contemporary media landscape, it reframes the problem of climate-crisis journalism from one of information transmission to one of publics and the challenges they face in the networked media environment. Moreover, it demonstrates that the climate crisis is not just a scientific phenomenon but also a social one and that the contemporary information crisis is driven not just by technological changes but also by concentrated and unaccountable power over our information infrastructures that long predates the rise of big tech. This unchecked power thrives in part because of the liberal tradition under which media operate in the United States, a tradition that champions above all the rights of individuals to speak (no matter how loud or toxic their message) over the collective rights that would ensure our right to hear one another and gain the benefit of a rich diversity of perspectives and voices. As I illustrate throughout the book, there is good reason to shift how we conceive of and regulate so-called freedom, especially when it comes to media.

Throughout the book, I use the terms *information crisis* and *information infrastructures* to refer broadly to the institutional and material dynamics that shape how publics connect and communicate. I use the terms *media landscape* and *media environment* interchangeably to refer more specifically to the impacts of these infrastructural-level dynamics on everyday mediated realities—things such as privacy, our ability to connect with one another, notions of truth, as well as the work and circulation of journalism and activist media. Just as the climate crisis is happening to the planet (and the lives it sustains), the information crisis is happening to the media landscape (and the connections it sustains).

The Mediated Climate builds on research that I have conducted over the past fifteen years on climate journalism, activism, and the rapidly changing media environment. I conducted much of this work as part of the MediaClimate research team, led by Elisabeth Eide at Oslo Metropolitan University and Risto Kunelius at the University of Helsinki and comprising scholars from around the world. We have done large-scale comparative analysis of coverage of the annual United Nations (UN) Conference of the Parties (COP) meetings and, more recently, the release of the IPCC reports.[13] Subsets of the MediaClimate group have collaborated on various other climate- and media-related research projects. Over the course of the past fifteen years, the group has expanded its focus from newspaper stories and journalists at legacy news outlets to activists and other civil society actors, scientists, and emerging online news outlets, with a keen focus on the dynamic between the changing media landscape and the changing role of these various actors and outlets in climate discourse. Countless conversations and research collaborations with MediaClimate colleagues are the inspiration for the questions raised throughout this book and for the answers I propose. Of course, any shortcomings are entirely my own.

As the climate crisis abides no boundaries, the book draws on research I have conducted on cases from around the world that shed light on the dynamic between the information crisis and the climate crisis. Analysis of the intersections where these crises meet, however, is far too big a task for one book. The political, economic, and media contexts that shape how the two crises are unfolding and how they are addressed differ greatly across nations and regions of the world. Because I am most familiar with the U.S. context, I focus most closely on it. I have interviewed journalists, editors, and activists in the United States over the past decade and have analyzed media use and content of U.S.-based journalists, activists, and scientists. The United States is a compelling case in part because of its extremes. It has some of the most pronounced political polarization as well as the most aggressive and well-funded campaigns of climate-change denial and cover-up. It is the center of the burgeoning tech industry and has some of the loosest tech regulation. It is also the site of robust journalism innovation and a hub of activism, advocacy, and climate-science expertise, which raises larger questions about the dynamics among journalism, science, and activism. My vantage point is specific, but it raises a series of straightforward but critical questions that can and, I believe, should be applied to other contexts.

One central aim of this book is to consider the state of contemporary publics. What kind of publics is our media landscape cultivating? What sort of publics would we prefer to be cultivating? And how can we begin to do this work?

Publics are constituted through communication. People need not consciously opt into membership in a public,[14] but sometimes they do so when inspired by a pressing problem that needs solving.[15] In perhaps the most idealized version, the public is a social formation of people identifying and working out a solution to a shared problem in the public sphere or one overarching sphere of discourse, accessible to all as the site of discussion of shared common interests and where everyone has equal ability.[16] In reality, though, access and ability to be heard are not equally distributed, so the overarching public sphere might be more accurately thought of as a space in which ideas and perspectives vie for visibility but where the playing field is vastly uneven. It is thus useful also to think in terms of counterpublic spheres, or smaller groups connecting with one another and the larger publics based on a group identity marker such as gender, race, sexual orientation, or ethnicity[17] or based on interest in a particular issue, often deliberately arguing against dominant perceptions of their group and of the interests and issues they represent.[18]

Indigenous water protectors, for example, who stand at the frontline of resistance to oil and gas pipeline construction are a counterpublic in the sense that they resist dominant notions that their interests can be discounted, they discuss common interests and goals among themselves, and then they collectively engage with so-called dominant publics through appeals to justice. Indigenous water protectors, along with the masses of protestors who have joined their efforts to stop gas and oil pipelines in Canada and the United States in recent years, can also be considered mobilized publics, or movements of people who collectively resist specific decisions, plans, and policies and call for specific solutions. In this case, water protectors resist the construction of oil and gas pipelines and call for respect for Indigenous communities and nature.

Publics are also constituted through communication in the sense that the public is a realm, a creative and culture-forming "space of appearance," where, as Hannah Arendt puts it, "I appear to others and others appear to me," where people act and speak together, wherever they may be.[19] For Arendt, the public realm is made of what she calls the "stories"—artifacts, relationships, and culture—that people create and hold in common. Water protectors, for example, have appeared to members not physically present at protests and to larger publics mostly through media—mobile group-messaging apps, social media feeds, and alternative and mainstream news outlets—as well as though artifacts of engagement or content circulated, from poetry to drone-camera footage.[20] See chapter 3 for more on this counterpublic.

Indeed, publics, whether considered social formations or spaces of appearance, depend on the infrastructures of public life because publicness at its core is always about connection. Benedict Anderson illustrates this in *Imagined Communities* (1983), where he argues that nineteenth-century journalism and other forms of print capitalism produced imagined publics. Daily journalism taught people to see themselves as members of cities, states, and national communities. Journalism fueled identity construction, Anderson and others have argued, by providing a foundation of shared information, values, and ritual.[21]

Since Anderson first detailed the role of journalism in constituting publics, the power of journalism to create imagined publics persists but in a diluted form because it has been tangled up in a larger media and information crisis. As digital networked infrastructures proliferate, trust in the press erodes, journalism economic models fail, and information-verification practices become muddled. Although digital communication technologies and platforms can fuel informed debate and decision making, they also pose new challenges to healthy public spheres and to the larger project of democratic society through intensified media business models, infrastructures that privilege profits over public good, and industry regulatory capture, where government agencies are staffed with people who support and often benefit from the industries they oversee.[22] Publics today, too, are fractured by receding shared standards for truth, which in turn create blurred lines between facts and so-called alternative-facts or falsehoods, knowledge and opinion, belief and truth. At the same time, though, with digital media there is also arguably more opportunity for visibility and influence among traditionally marginalized groups and perspectives.

In the roughly forty years since fossil-fuel-generated climate change was widely acknowledged as a threat, there has been much investigation into why forces of climate denialism, skepticism, and contrarianism remain so strong and, more broadly, why efforts to respond to the climate crisis lack power and effectiveness.[23] Many critics still blame journalism for failing to cover the topic in urgent and compelling ways. "The media are complacent while the world burns," read a piece in the *Columbia Journalism Review* in 2019. "Instead of sleepwalking us toward disaster," wrote authors Mark Hertsgaard and Kyle Pope, "the U.S. news media need to remember their . . . responsibilities—to awaken, inform, and rouse the people to action."[24]

Such criticism is based on the assumption that public ignorance, apathy, and denial are evidence of news-media shortcomings. This view has spurred important related research and discussion and has prompted newsrooms to reassess their approaches to climate coverage. But the assumption ignores the fact that journalism has been tangled up in long-standing tensions between market interest and public interest that have been intensified in the context of today's media landscape, where even the best work must fight for visibility and compete in an environment that is quite literally gamed against quality content and serious debate.

This is why transforming the media landscape must be an integral part of the effort to address the climate crisis. Toward that end, *The Mediated Climate* explores the climate crisis as a news story—one of the most important news stories in history and surely the most pressing news story ever to be successfully distorted, disputed, diminished, and diluted over the course of decades. The story of the climate crisis as news story has much to tell us about the contemporary media landscape, the obstacles it presents to healthy public discourse, and how these obstacles might be overcome. The book offers an analysis of the role played by media corporations, journalists and other media professionals, tech companies, and activists in shaping the story of the climate crisis. The analysis considers historical forces, including the long-running tension in U.S. journalism and the larger public discourse in the U.S. public sphere between corporate interests and public interest and how that push-and-pull has slowed effective action on accepting and addressing our climate reality.

Central to my exploration is an attempt to examine the blurred boundaries and interplay among politics, journalism, science, technology, public relations, and activism and to better understand the dynamics between information and affect—all part of a media environment marked by shifts in the balance of power that shape information production.[25] As someone who has studied the evolving networked media landscape since the mid-1990s, I have seen firsthand and documented how these blurred boundaries create tensions, for example, between top-down control and distributed networks, between manipulation and informed engagement, and between mass surveillance and sophisticated privacy protections.

The internet space is marked by contradictions and competing understandings of what it is and ought to be. In the mid-1990s, for example, the international collection of emailers and listserv participants that formed around the Mexican Zapatista revolutionary movement against transnational free-market capitalism created and distributed alternative and mainstream media coverage of rebel clashes with the government.[26] To them, the internet was the means to create a more active public that could scrutinize information and counter disinformation in real time and a place where they could write and publish their own critiques, updates, reports, and opinions. It was a counterforce to the isolation of dissidents and outsiders that had been a product of media marginalization and that had minimized effectiveness of activist movements in the past.[27] In the 1990s, just as the business model of newsrooms collapsed, the Zapatistas led the charge of less professionalized newsmakers whose bottom line was not making a profit but rather getting a seat at the table of public conversation. These early "netizens" initiated a trend toward alternative digital news networks that stemmed first from activist movements and later from groups not necessarily associated with specific causes or movements. In 1999, the Seattle Independent Media Center, a collective of independent media organizations and hundreds of journalists offering what they described as grassroots noncorporate coverage, used the internet to support and coordinate protests by amateur or "citizen journalists" at World Trade Organization meetings.[28] These early conceptions of the internet as a democratic space—a freer alternative to mass-mediated space—that could support more pluralistic discourse have unfortunately since that time been joined by much more noxious uses of the internet in which users are surveilled, information is weaponized and distorted, and the architecture is highly controlled.

Scholars, including me, have written extensively on how and to what end digital platforms now host our political, commercial, professional, and social interactions. There is, however, little work that specifically connects the changes in journalism information practices and infrastructures to the ways the climate debate has played out. To fill that gap, *The Mediated Climate* traces the history of the way the climate crisis and the information crisis intersect. My aim is to highlight how the fundamental workings not just of journalism but also of the larger media landscape have shaped and continue to shape the discourse and policy formulation around the climate crisis. By blending theoretical and conceptual discussion with detailed analysis of key characteristics of the "hybrid media system," including misinformation, data, and activism, the book lays bare the power dynamics at work behind climate discourse, including the challenges to and possibilities of disrupting the current system.[29] It stands at the intersection of academic analysis and solution-driven critique in large part because writing a book about these crises demands an exploration of not just what has gone wrong but also what we can and ought to do about it. Hope must be part of that exploration. We are at a crucial moment, one of exciting possibilities, driven first and foremost by the fact that the status quo is untenable. *The Mediated Climate* should be read as a provocation to consider the media and natural environment in tandem, so we might get rid of the pollution choking both of them.

Although my focus here is on how our information environment shapes how we think about and act on the climate crisis, I must also point out that the companies that serve as the backbone of the internet are not only polluting the public sphere but quite literally polluting the environment, too.[30] The industry uses massive amounts of electricity to store data and train artificial intelligence (AI), and the major players in big tech— Alphabet (Google's parent company), Microsoft, and Amazon— all contribute to the climate crisis through billions of dollars of contracts with big oil, using automation, AI, and big-data services to make oil exploration, extraction, and production more efficient.[31] Until 2020, Google had an energy division that was created specifically to win the business of oil and gas companies.[32] All these practices accelerate the climate crisis.[33] At the same time, though, big-tech companies publicly promote themselves as pro-climate, making bold climate pledges and supporting tech fixes such as carbon capture.[34] They are, however, conspicuously absent as advocates for stronger climate policies; Apple, Amazon, Alphabet, Meta, and Microsoft spent roughly $65 million on lobbying in 2020, but only 6 percent of that huge amount targeted climate policy.[35]

WHAT WE KNOW ABOUT COMMUNICATING ABOUT THE CLIMATE

A rich body of work across various fields, including journalism studies, science and environmental communication studies, strategic communication, political science, and media studies, tells us that understanding climate communications—what shapes the discourse and the types of stories that are effective at spurring public understanding and engagement—is exceptionally complex. This is because of the dynamics among the actors involved; the high stakes of both acknowledging and not acknowledging the gravity of our collective situation; and the phenomenon of climate change, which involves science that is difficult to understand and seems distant and abstract to those who have not yet experienced the climate crisis firsthand—although they are a rapidly dwindling portion of the population. Practices are evolving to help more effectively communicate about the climate crisis.

Understanding the media effects on our willingness and ability to do something about the climate crisis is also complex because even after nearly one hundred years the field of communication can say little about such media effects except "it depends." This is not a failure of the field but rather a problem that stems from the complexity of the phenomenon. We know, for example, that professional practices of journalists and their sources shape the story. We know that the message sent is not always the message received, that understandings of both narratives and facts very much depend on what we called in the mass-media era the *receivers* or *audience* but what is more often referred to today as the *users* to reflect the active role they play in the communicative process. The cultural and political contexts in which receivers or users exist, their life experiences, their preferences, their moods, and so on can affect interpretation. We know that the amount and prominence and type of media representation often steer our collective attention toward some things and away from others, a process that Bernard C. Cohen famously called "agenda setting." That is, news media "may not be successful in telling its readers what to think, but it is stunningly successful in telling its readers what to think about."[36] The ubiquitous adoption of networked and mobile technologies makes clear the impact that media tools—television, mobile phones, social media platforms—have on how and what we communicate. This is what Marshall McLuhan famously argued long before the advent of the internet: the medium is the message. A popular digital-era version of this argument among scholars contends that the medium, with its material affordances, is always part of the message.[37] We know that institutional and material conditions—or media infrastructures and tools—have an enormous influence on how we engage and understand one another and the world. To assess the power of media to shape public understanding and action—or misunderstanding and inaction—around the climate crisis, it's worth considering all these facets—practice, audience, representations, and institutional and material conditions—as well as the inherent ambiguity of media's impact.

PROFESSIONAL PRACTICES

Journalism

When we explore the question of our collective failure to communicate effectively about the environment and facilitate climate action, we must consider the workings of both journalists and scientists, who share central epistemological tenets. Journalism in fact has long borrowed from science in its commitment to objectivity, and that commitment has been failing them both when it comes to climate-crisis communication.[38]

Objectivity as a goal in news reporting came with the invention of the telegraph in the early 1840s and the subsequent birth in 1848 of the first American wire service, the Associated Press, and other wire services attempting to produce reporting acceptable to the politically varied papers that they served. Striving for objectivity did not become the norm, however, until after World War I. Wartime propaganda and public-relations campaigns convinced journalists that facts could not be trusted, that what they reported was too often created for them by interested parties. In response, reporters adopted rules and procedures alleged to result in objective reporting in order to signal the legitimacy and high standards of their work.[39] Journalists' growing professional faith generated social cohesion and occupational pride, on the one hand, and internal social control, on the other. It was during this high-modern period of ideological consensus—from the end of World War II until roughly the 1980s—that the norm of objectivity developed, especially in the United States. However, with the breakdown of ideological consensus in the 1960s and onward—by the Vietnam War; by conflict over race, gender, and sexuality; and by hyper free-market economic policy—practices associated with objectivity have become increasingly recognized as creating their own biases.

The professional norms of mass-media journalism attempted to use objectivity to separate facts from values or opinions, but the journalism scholar Theodore Glasser writes that these mechanisms for creating and maintaining truth create a predictable series of biases. Practices of objectivity favor the status quo because they encourage reporters to rely on bureaucratically credible sources. They discourage independent thinking because they dictate that journalists are mere spectators, compelling them to attempt to leave their own inevitable opinions and insights out of their stories. And they are biased against the idea of responsibility: the objectivity that journalists claim exists in their stories absolves them of accountability for the content.[40]

We can see clearly how this objectivity/bias problem played out in climate-crisis reporting. Max Boykoff and Jules Boykoff's landmark study in 2004 found that U.S. journalists covering climate issues for legacy news outlets under the guise of objectivity created a "faux balance" or false equivalence, giving equal weight to both scientific evidence of the existence of anthropogenic climate change and so-called climate deniers who doubted its existence. That is, by uncritically relying upon and conveying the point of view of bureaucratically credible sources, namely scientists and politicians, reporters produced an unrealistic picture of the validity and prominence of deniers and in some instances rewarded the dishonesty of political and corporate elites who relied on reporters' objectivity to amplify rather than question their lies.[41] Chapter 1 delves more deeply into the ways journalistic attempts at objectivity through balance provided a space for massive corporate misinformation campaigns and stalled meaningful widespread environmental discourse and policy. Suffice it to say for now that when it comes to reporting on the climate, the bias carted along with objectivity has resulted in a vastly inaccurate view of scientific knowledge.

In recent years, some environmental journalists and media outlets have had a sort of reckoning and have been updating their approaches to covering the climate crisis. Michael Brüggemann refers to a "post-normal" information landscape in which journalists are replacing the professional standard of objectivity with standards such as fairness and transparency in order to create spaces for greater engagement while prioritizing a dedication to verifiable facts over neutrality.[42] For environmental journalists, this shift in approach has meant refusing to place climate research presented by specious figures on the same level as research delivered by scientists legitimated by the scientific community. Moreover, it entails amplifying the voices of authorities who use the scientific consensus on the climate crisis as a starting point for their positions.[43]

Such organizational and editorial policies, combined with the introduction of new digital players and the changed business model of journalism, are reshaping how the climate crisis is covered. Online outlets such as *Vice*, *Buzzfeed*, and the *Huffington Post* have dedicated resources to delivering environmental coverage that is more engaged with climate-justice efforts than was past climate journalism.[44] In addition, a University of Oxford Reuters Institute study on how people access news about the climate crisis in forty different countries reported in 2020 that 13 percent of people surveyed say they pay attention to niche or specialized sites covering climate issues. Indeed, the number of climate- and environment-specific sites is on the rise, some of which enjoy a wide readership and financial success.[45]

In addition to rethinking the practices of science reporting, the work of climate journalists is interrelated with the work of environmental nongovernmental organizations (NGOs). Press releases by NGOs, for example, are strong and systematic drivers of media attention to the climate crisis.[46] And especially during transnational events such as the annual UN climate-change conferences, the professional boundaries of the complicated and often antagonistic relationship between journalists and public-relations professionals have blurred in part because of the complexity of the issues at hand and the "camp feeling" that develops when reporters and NGO communication professionals are in the same space and therefore have more opportunity to interact.[47] Some news outlets even partner directly with advocacy groups, as is the case between *The Guardian* and 350.org.[48]

These varied voices and partnerships are in part caused by disruptions to the business model of newsrooms, especially among legacy news media, which fund the majority of professional journalism and for whom traditional revenue sources are declining.[49] In order to adapt to new attention infrastructures and to confront widespread disinformation, the press has entered into partnerships with social media platforms. Beyond these institutional arrangements, the role that media tools, platforms, and algorithms play in shaping our media environment and how we act within it demands consideration.

Science

The epistemic structures and practices of Western science have also gotten in the way of effective climate communication and action. As in the case of journalists, scientists' commitment to professional practices related to objectivity and to demonstrating the legitimacy of their work has made them reluctant not only to advocate for solutions to the problems unveiled through their work but also even to make predictions—which is the only way to communicate the impact of the climate crisis and other environmental changes over time.

Scientists develop in-depth expertise in narrow areas—fish biology, soil science, oceanography, for example—to ensure high standards and robust understanding, which makes it hard to investigate complex systems such as climate change. Even scientists who take a broad view often feel it would be inappropriate to articulate what they see because doing so would demand speaking beyond their area of expertise. We can see this reluctance even in reports from the IPCC, a UN body that evaluates and synthesizes climate science in reports every five to seven years. The IPCC involves hundreds of scientists and draws on the work of thousands more from various fields. Its goal is "to ensure an objective and complete assessment and to reflect a diverse range of views and expertise."[50] Because of this commitment to objectivity and focus on natural systems, social science perspectives were largely absent from the first three assessments in 1992, 1999, and 2001 and addressed in only limited ways in the fourth and fifth assessments in 2007 and 2014–2015.[51] It was not until the sixth report, released in 2022, that there was an entire chapter dedicated to the social aspects of climate mitigation, highlighting among other things the impact of continued fossil-fuel dependency. "It's now or never," said Jim Skea, a cochair of the report. "Without immediate and deep emissions reductions across all sectors, it [mitigation] will be impossible."[52]

In addition to niche expertise and avoidance of discussion of social causes, excessively stringent standards for accepting claims—standards rooted in scientists' "desire to demonstrate their disciplinary severity"[53]—also pose a challenge to clear communication about climate change. Among scientists across disciplines, it is thought to be far worse to fool oneself into believing something that doesn't exist (a type I error) than not to believe something that does (a type II error). Protocols are developed to avoid type I errors at all costs. And thus it is far more professionally risky to predict the impacts of the climate crisis than to remain silent.[54]

These structures and practices of Western science have left climate scientists and their research particularly vulnerable to misinterpretation and sabotage. Just as faux balance in journalism amplified the lies of climate deniers, the tenets of scientific professionalism that make scientists reluctant to advocate solutions to the problems identified by their work also make them vulnerable. Political and corporate entities whose interests are threatened by recognition of the climate crisis further their campaigns to promote denialism by harassing and threatening scientists and spending large sums in effort to undermine scientists' credibility and exacerbate public mistrust.

Candis Callison argues in her book *How Climate Change Comes to Matter: The Communal Life of Facts* (2014) that scientific fact must be perceived to be created with complete neutrality but also to be loaded with meaning once it leaves the scientific context.[55] This tension between neutrality and engagement creates a double bind for those invested in understanding climate and other science communication. On the one hand, climate change is a combination of scientific facts based on empirical observation as well as on modeling and projection. Accepting the assertion that the climate is changing in an adverse way, then, requires faith and trust in science. On the other hand, to engage people and to sort out what to do, climate change must become about more than cold, hard facts. Climate change, then, says Callison, "must promiscuously inhabit the spaces of ethics, morality, and other community-specific rationales for action while resting on scientific methodology and institutions that prize detachment from politics, religion and culture."[56]

This double bind describes the central tension behind a rethinking of objectivity in both science and journalism. Both professions have increasingly recognized that detachment and the democratic obligation to inform and engage publics are at odds with each other.[57] Both professions also rely on an information-deficit model, which assumes that a higher level of public knowledge will result in support for and interest in science.[58] Put another way, they function on the premise that more and better information is the solution to ignorance and inaction when it comes to the climate crisis and other problems. Despite the well-documented adverse impact of false information, however, more and better information doesn't necessarily lead to awareness or engagement. The dynamics between facts and changes in public attitudes about the climate crisis are complex, and the information-deficit model can mask the underlying purposefulness with which environmental degradation persists. This issue is explored further in chapter 1.

REPRESENTATIONS

Just as journalists and scientists tend to focus on the neutrality of their depictions of the world, communication researchers tend to focus on content, or the power of representations. In the 1920s, Walter Lippmann proclaimed the power of mass media to "put pictures in our heads." These pictures, he argued, did not represent the world as it is but rather the interests of those who create them. Lippmann explained that it is the content creators (or their financial and political backers) who fundamentally shape the pictures presented. His central concern was how newspapers could inform those pictures by winning legitimacy through identifiable editorial norms and practices.[59] The media scholar Fred Turner points out that Lippmann was espousing an idea—media's power is rooted in the ability to create and distribute representations and through them to shape perceptions—that has been a central belief in media theory for the past century. Audiences can be overpowered by these pictures or more active makers of meaning. Either way, representations and the worldviews carted along with them are seen as constituting the central vein of media power in the mass-media era.[60] This emphasis on the power of representation endures today—thus, with respect to the climate crisis, much of the focus is on how climate is represented and how it might be represented to better effect.

One prominent mode of representing the climate crisis, is, of course, through data. Measurements of global temperature rise, warming and rising oceans, shrinking ice sheets, glacial retreat, and ocean acidification serve as stark indicators of the rate and severity of the crisis. But data are cold.

Perhaps the most agreed upon point about climate communication is that stories are more effective than data. For example, a study by the political scientists Deborah Guber, Jeremiah Bohr, and Riley Dunlap in 2020 analyzed floor speeches in the U.S. Congress from 1996 through 2015 and found that Democrats most often made arguments about climate change backed up by facts and data, whereas in denying climate change Republicans tended to tell stories and use imagery, emotional appeals, and humor to sway people to their side. Republicans, they argue, are communicating in ways that may ultimately be more effective.[61] In one famous example, on a cold winter's day in 2015 the Republican senator James Inhofe of Oklahoma, a former real estate and insurance executive, brought a snowball to the Senate floor, presenting it as proof that global warming was a hoax. He held it up and asked: "You know what this is? It's a snowball, just from outside here. So it's very, very, cold out. Very unseasonable." He then tossed it to Senator Bill Cassidy of Louisiana, who was presiding over the Senate's debate, saying, "Catch this!" Inhofe, who at the time was the chairman of the Senate Committee on Environment and Public Works, was widely ridiculed for the stunt as idiotic, but it got a lot of attention. Guber and her coauthors point to the stunt as "a prime example of using that vivid imagery to communicate something about climate change that certainly isn't true, but the truth of it doesn't really matter to the audience."[62]

One of the devices used to turn facts into narratives is the news frame, which creates cognitive maps that people use to organize their understanding of reality.[63] According to Robert Entman, frames "select some aspects of a perceived reality and make them more salient in a communicating text, in such a way as to promote a particular problem definition, causal interpretation, moral evaluation, and/or treatment recommendation for the item described."[64] Studying different climate-related frames and people's responses can allow more tailored and thus more relevant communication, better conceptual boundary making, and clearer connections between concepts and understandings.[65]

Climate communicators, including scientists, journalists, and advocacy professionals, strive to find frames that resonate more effectively with audiences. A handbook for IPCC authors published in 2019 and meant to guide scientists in effective communication about the report suggests, for example, that avoiding wastefulness has particularly strong resonance with conservative audiences and that health-benefit frames—for instance, emphasizing the advantage of clean air, less traffic, and more cycling and pedestrian traffic on the road—are generally well received by a cross-section of the population.[66]

So-called disaster or apocalypse frames are frequently used but also hotly contested by those who fear their alienating impact.[67] Such frames tend to paint a picture of a climate crisis so alarmingly advanced that either there is nothing we can do, or only extreme measures will do. For example, David Wallace-Wells's *New York Magazine* cover story "The Uninhabitable Earth" in 2017, the most widely read story in the magazine's fifty-plus-year history, opens with this line: "It is, I promise, worse than you think. If your anxiety about global warming is dominated by fears of sea-level rise, you are barely scratching the surface of what terrors are possible, even within the lifetime of a teenager today."[68] The piece and a subsequent book of the same title explore worst-case scenarios, but some climate scientists criticized them for overstating the extremity of future scenarios or, as the glaciologist Peter Neff put it, for taking "significant literary license to leverage information grounded in truth and paint an apocalyptic picture of extreme future scenarios."[69] But for Wallace-Wells, "There is no single way to best tell the story of climate change, no single rhetorical approach likely to work on a given audience, and none too dangerous to try. Any story that sticks is a good one."[70]

Clearly, different audiences respond to different rhetorical approaches. But Wallace-Wells is wrong: any story that sticks is *not* necessarily a good one if by "good" we mean accurate, not merely attention grabbing. Consider again Inhofe's snowball. From the perspective of effective climate communication, certainly there are better and worse ways to tell the climate-change story. One particularly problematic subgenre of the disaster frame written by the so-called sad bois is superniche but case in point.

The "sad bois" framing of the climate crisis is the product of novelists who have turned their attention to writing essays on the climate crisis: Jonathan Franzen, who wrote in the *New Yorker* on the sorry future of birds; Michael Chabon, whose *Paris Review* piece laments his own lack of impact; Nathaniel Rich, who detailed the end of climate policy making in the *New York Times Magazine*; and Jonathan Safran Foer, whose book *We Are the Weather: Saving the Planet Begins at Breakfast* argues that we not eat meat until dinner. In a scathing critique of the genre, the journalist Kate Aronoff points out that "reading Foer, Franzen, and the other novelists turned climate catastrophists brings up the question 'Who, for them, is "we"?' The Global North has historically fueled the climate crisis, while the Global South is experiencing its effects now, as with catastrophic flooding in Bangladesh." Yet the sad bois' tendency to display the climate crisis through the lens of their own existential demise and to reduce climate solutions to personal morality leads to absolution rather than to economic and political reform. Go vegan. Get a bike. Take up birdwatching. Aronoff comments, "Far-off climate disasters are mentioned only briefly in Foer's book, and if they appear in other doomist books and essays, it is mainly as tragic set pieces." The sad bois don't seem to want to see reform but rather only personal absolution. Indeed, as Aronoff further points out, "If the world does manage to steer away from catastrophe, the credit will be owed to a critical mass of social movements, unions, and the elected officials accountable to them, working to take power back. No angst-filled breakfast or lunch can do the same."[71]

Michael Mann, a climatologist and geophysicist whose groundbreaking work contributed to the scientific understanding of historic climate change based on the temperature record of the past thousand years, points out how blaming the individual is not just a misguided approach but also a strategy used to deflect accountability from the industries responsible for harm: "For a UK example look at BP, which gave us the world's first individual carbon footprint calculator. Why did they do that? Because BP wanted us looking at our carbon footprint not theirs."[72]

The individual-responsibility frame promoted by the sad bois and others runs directly counter to the climate- or environmental-justice frame, which connects structural inequities to environmental degradation. The latter frame includes discussions of historic responsibilities for the climate crisis, calls for a halt to unfair and disproportionate burdens imposed on different countries and communities, such as the working class and communities of color. It is increasingly prominent and overlaps with an increase in coverage of the climate-justice movement.[73] For example, a study of coverage of the Paris UN

Climate Summit in 2015 documents voices from activists and NGOs in both legacy media outlets and online players such as *Buzzfeed*, *Vice*, and the *Huffington Post* in the United States. The environmental-justice frame was taken up more prominently by *Vice*, where it was mentioned in more than half of all articles. *Vice* also devoted by far the most attention to covering protests and rallies, which appeared on average in every second *Vice* article.[74]

Milane Larsson, a senior producer and reporter at *Vice*, points to how discussions of contested justice issues enable her to "take people on the journey" and provide "immersive pieces in the crowds" at protest rallies.[75] In a shift from typical coverage of protests and social movements, which emphasizes conflict and chaos over issues and proposed solutions, the environmental-justice frame provides a space to address the structural problems that exacerbate climate injustice, including unchecked environmental degradation, increased health and safety effects in economically disadvantaged countries and communities, and climate-driven mass migration.

Ultimately, this tension between personal and structural framing of problems points back to the fact that although the changing climate is a scientific phenomenon, how to respond to it is a question of values, and understanding people's values is a key component to effective climate communication.

AUDIENCE

We know that when it comes to contested issues such as the climate crisis, people's opinions are consistent with their "cultural way of life," aligned with how they think things should be and with what they perceive to be their peers' attitudes.[76] Believing (or not) that we are in the midst of a climate crisis, that is, has more to do with political identity than with logic.

Linking a story to people's already-existing ethics and understanding of the world and their communities plays a central role in public engagement with complex scientific issues.[77] In other words, people ascribe a different meaning to climate change depending on their already-existing belief system.

If you are a parent, you may believe you have an obligation to your children; if you are a business leader, you may care about climate change to show your commitment to social responsibility; if you are religious, you may feel protective of the natural world based on that particular belief system. Katharine Hayhoe, an atmospheric scientist who is celebrated as "one of the nation's most effective communicators on the threat of climate change and the need for action," argues for the need to talk about why climate change matters to us in the places where we live. "If we do that," she says, "we will find a way to get past the political polarization linked to the phrase climate change."[78] Max Boykoff similarly argues that the key to effective climate communication is knowing your audience, valuing their points of view, and emphasizing how climate change affects their everyday lives. "This work to enhance relevancy," he writes, "then creates openings for discussion and action."[79]

These calls to listen are not meant just to connect with people who don't see the relevance of or believe the reality of climate change. They are also a central feature of climate-justice efforts, whereby journalists and policy makers create time and space for communities most affected by environmental degradation to voice their concerns and to be heard.

As a strategy to achieve climate justice, however, listening can seem a bit feeble in the context of larger systemic forces. Corporations pump billions of dollars into lobbying the government and producing and spreading disinformation. Tech platforms privilege profit over public interest and have a huge and often unseen sway over climate communication. Indeed, the power dynamics are so stacked against the communities most affected by the climate crisis that listening hardly seems enough to correct for it.

Chenjerai Kumanyika commented on Twitter about the attempts by white America to listen in order to come to terms with its racism in response to the Black Lives Matter movement of 2020:

> The idea of leaders "listening" or having a sit-down also invites us into the ahistorical delusion that injustice results from the ignorance of leaders (rather than their priorities and class interests) and that transformative change happens because leaders willingly make it so. Whatever rights POC's [peoples of color], women, queer folks, people with disabilities have earned[, we] didn't earn those because leaders learned through dialogue what was wrong. Instead, the historical record is clear. Dominant institutions were FORCED to do things they didn't want to do. Of course in the broader sense, "listening" is important. But listening in the context of struggle or in what is actually political negotiation is completely different from these depoliticizing, corporate "listening" strategies.[80]

The same can be said for leaders "listening" to communities who bear the brunt of climate injustice. How can we "listen in the context of struggle" with the full acknowledgment that the climate crisis and associated injustices are not the result of ignorance but of greed, cover-ups, intentional noise, and willful disregard for public interest?

In their book *You Are Here* (2021), Whitney Phillips and Ryan Milner detail the ways in which "polluted info is as damaging as it is perfectly calibrated." Like climate change, "networked change emerged slowly over time, as existing filtration systems became more and more taxed by pollution."[81] As in the pollution of information, environmental degradation is simply not a malfunction in the system. In this system, wealthy nations propel infinite growth and expansion and never-ending mass production and consumption, while economically disadvantaged nations and individuals bear the brunt of environmental degradation.

Indeed, listening in the context of struggle—whether it be to climate deniers or to communities whose air is choked by industrial pollution—must involve a critique of the power structures and media practices that got us where we are today. As Phillips and Milner point out, "The issue is how the world around us brought us to this moment and how it reflects a system working exactly as intended."[82]

INSTITUTIONAL AND MATERIAL DYNAMICS

As has already been established, the climate crisis and the information crisis we currently face intersect, each exacerbating the other. In considering the ways in which they intersect and the debilitating amount of "discursive pollution" in our media landscape, it is worth noting that online environments, including the platforms and tools that shape them, are constructed by people and within economic, social, and political frameworks but that natural environments are not.[83] As Phillips argues, "A redwood forest couldn't be otherwise. A digital forest could be (and almost was)."[84]

Our tools and infrastructures embody dominant ideologies and the norms that we apply to govern them. As Geoffrey Bowker and Susan Leigh Star describe, infrastructures are scaffolding that structure the conduct of modern societies.[85] They constitute relations between materials and practices. When working, they are embedded, standardized, and largely invisible. When they break down, they become obvious.[86] If we miss this starting point—that our tools, technologies, and the practices they favor could be otherwise—the inequalities and biases that they reproduce seem inevitable or natural. Indeed, the chaos that characterizes our current information landscape is not a glitch in the system but rather a constitutive feature of it. The networks we use to communicate facilitate what Frank Pasquale calls an "automated public sphere,"[87] in which platforms such as Facebook and Google have come to control the type and quality of content, turning engagement into data points, ordering content based on market considerations rather than on public interest, and facilitating the "tyranny of the loudest," a phrase coined by the *Buzzfeed* reporter Lam Vo.[88]

Tech leaders routinely sidestep any acknowledgment that values are embedded in design. Facebook's CEO, Mark Zuckerberg, consistently claims that his companies are technology rather than media enterprises. In doing so, he gives the impression that technology is neutral. From that standpoint, "Facebook can claim that any perceived errors in Trending Topics or News Feed products are the result of algorithms that need tweaking, artificial intelligence that needs more training data, or reflections of users," as Mike Ananny, a University of Southern California communication professor, puts it. "Facebook [can] claim that it is not taking any editorial position."[89] Zuckerberg and other Meta spokespeople do this in part because categorizing Facebook as a media company would raise questions about the impact of liberalism (or lack thereof) at the heart of internet governance. We'll return to that specific issue later, but first let's consider the ways that our built environment, whether physical or informational, is run on tools and infrastructures imbued with political and social values.

A particularly pernicious feature of the contemporary information environment is algorithms that sort content and exert profound sway on our social, political, and economic realities. Examples are well documented. In *Algorithms of Oppression* (2018), Safiya Noble demonstrates the way biases in search algorithms privilege whiteness and discriminate against people of color, specifically women of color. In 2009, she typed "Black girl," "Latina girl," "Asian girl" into the Google search engine, and the first page of results invariably listed pornography sites, challenging the idea that search engines such as Google offer an equal playing field for all forms of people, ideas, identities, and activities.[90] Joachim Allgaier's research documents the way YouTube algorithms boost controversial and extreme points of view to create conditions that privilege climate denialism. Search "global warming" on YouTube, and you are likely to find results that oppose scientific consensus or that characterize the scientific debate on climate change as significant and ongoing when there really is no debate.[91]

In an interview by Michael Barbaro, host of *The Daily* podcast for the *New York Times*, just before the 2020 U.S. presidential election, Twitter cofounder and then CEO Jack Dorsey acknowledged how human decisions drive algorithms to perform in particular ways and lamented not hiring experts to help him understand the impact of small design choices. "The disciplines that we were lacking in the company in the early days, that I wish we would have understood and hired for," he said, were "a game theorist to just really understand the ramifications of tiny decisions that we make, such as what happens with retweet versus retweet with comment and what happens when you put a count next to a like button?" Without this expertise, he says, the company unwittingly built incentives into the app that encouraged users and media outlets to write tweets and headlines that appealed to sensationalism instead of to accuracy. He went on to describe how the platform's algorithm made it so that "the most salacious or controversial tweets will naturally rise to the top because those are the things that people naturally click on or share without thinking about it."[92]

This admission is impressive for its acknowledgment that Twitter was built, however inadvertently, to propel the tyranny of the loudest. But Dorsey's description of the rise of salacious and controversial tweets and the many clicks on them as *natural* misses the point and contradicts his previous admission that a design choice created this undesirable feature and that a design choice could undo it. Dorsey went on to suggest: "We need to open up and be transparent around how our algorithms work and how they're used, and maybe even enable people to choose their own algorithms to rank the content or to create their own algorithms, to rank it. To be that open, I think, would be pretty incredible."[93]

This proposed response—a choose-your-own-algorithm option—is a perfect example of how values become embedded in the tools we use to interact. It ascribes individual-level solutions to system-wide problems, a trademark characteristic of the liberal tradition under which media in the United States operate. *Liberal* here means neither politically progressive nor morally lax, as it is used in contemporary U.S. politics, but rather refers to the apolitical philosophy at play there. Although a broad spectrum of flavors of liberalism encompasses a range of political perspectives, in the broadest sense liberalism enshrines individual freedoms such as free speech, a free press, property rights, and civil liberties as well as a free market, free trade, and other hallmarks of decentralized government.[94]

Liberalism has had profound influence over both contemporary and historic information infrastructures. In the case of the internet, the mantra "information wants to be free" directly reflects libertarian ideals and the resolute defense of negative personal freedoms: freedom *from*. John Perry Barlow, cofounder of the Electronic Frontier Foundation, affirmed the negative freedoms of online spaces in his essay "A Declaration of the Independence of Cyberspace" (1996): "Governments of the Industrial World, you weary giants of flesh and steel, I come from Cyberspace, the new home of Mind. On behalf of the future, I ask you of the past to leave us alone. You are not welcome among us. You have no sovereignty where we gather."[95] These libertarian sentiments reflect the dominant thinking of early and contemporary internet architects. According to Phillips and Milner, "All these negative freedoms fundamentally shaped the digital landscape. First, freedom from censorship ensured that the maximum amount of information—regardless of how harmful, dehumanizing, or false—roared across the landscape. Freedom from regulation encouraged what journalism professor Meredith Broussard calls 'technochauvinism,' the overall sense that if something can be done, that's reason enough to do it. Build the website. Share the information. Thus spoke Mark Zuckerberg: move fast and break things."[96]

Negative freedom is built around the premise that regulation, particularly around content, is unnecessary because the marketplace of ideas will ensure that quality information will emerge to the top in the competition of ideas and gain widespread acceptance. By extension, this libertarian view advocates a hands-off approach to media regulation. This idea is widely embraced by the tech industry, but its efficacy is widely disproved by the bunk information that enjoys widespread circulation and popularity on its platforms. For platforms, maximized free speech leads to maximized profits, "so they [platforms] have twice the incentive to leave moderation to the marketplace of ideas."[97] In fact, Dorsey's appearance on *The Daily* was meant to assure the public and policy makers that Twitter would step up efforts to prevent being used to target voters with false information about election participation, especially in light of the fact that then president Donald Trump had already posted a series of lies about mail-in voting and declared that "this will be a rigged election." Indeed, the interview was part of a thinly veiled effort to assure us that no government regulation is necessary because Dorsey and Twitter had things under control.

In this exchange with Barbaro, Dorsey tried to argue convincingly that while the company had so far not been "awesome," with a strong feedback loop it could "do things right."

DORSEY: The one skill I want us to be incredible at, the one skill I want us to build is our capacity to learn. It's this cycle of observe, learn, improve. If we can be incredible at that cycle, I'm confident we'll do the right things no matter what challenges we're facing. . . .

BARBARO: Would you agree (this might seem harsh) that you have not been incredible about that? And maybe not even especially good at it?

DORSEY: About learning? I would agree that we have not been awesome, but I think we are getting better and better every single day and I think that is on display publicly, especially in this past year, around everything that we've learned and how we've evolved, our policies have evolved our actions and enforcement. The transparency the company has with the world now is unique and something I'm very, very proud of and goes much farther than most.[98]

Dorsey throughout the conversation holds up transparency as evidence of the company's worthiness to play host to large swaths of our public's discourse. Although transparency is all well and fine, he situates the power and responsibility for it on the company and the individuals in charge rather than on an agreed-upon set of standard and norms while at the same time upholding a dedication to *freedom from* or negative freedoms.

Liberalism's impact is not limited to big tech and has long had profound implications for the press. In *Networked Press Freedom* (2018), Ananny argues that the emphasis on *freedom from* embedded in our idea of press freedom has had profound effects on the public's ability to fully engage. Negative freedoms, he argues, hold that "if journalists and publishers can get truth to the public, then individual readers and viewers will be able to make informed decisions about how to think and vote."[99] But this obviously extremely limited view privileges individual right to speak over the public's right to hear the rich combination of voices and perspectives on which democracy ought to be built.

Dorsey's interview reveals the way the tech industry's celebration of individual solutions distracts us from our ability to consider our collective need and our ability to find more structural solutions. Ananny and others suggest that we need to be asking, What are the "deep structures" making some publics more possible than others? That is precisely what this book sets out to do.

STRUCTURE OF THE BOOK

The pollution in today's media environment obstructs effective communication about climate. Throughout the book, I detail the nature of the pollution and the efforts people can make and are making to clean it up. Each chapter addresses the role of

INTRODUCTION: TWO CRISES ⌘ 37

different forces in shaping climate communication and, in turn, climate publics.

Chapter 1, "House on Fire," explores the mechanisms of culturally produced ignorance around climate change; the roles that media corporations and the press play in creating and, in the case of journalism especially, attempting to combat that ignorance; and the environment that breeds the ignorance. After reviewing the history of efforts to undercut climate science, the chapter details the evolution of climate journalism and the profession's collective push to improve coverage of the climate crisis by reflecting on mistakes of the past and adjusting its practices, striving to deliver more and better coverage with the assumption that doing so will spur more healthy and productive climate discourse. Chapter 2, "Noise, Incivility, and Ambivalence," considers what is overlooked by the information-deficit model—the thinking that more and better coverage is a key to climate solutions—by examining the social and material contexts that affect both journalism and publics. After providing background on the historical relationship between publics and media infrastructures and the changing dynamics of this relationship, it presents the larger context in which journalism operates—the tools and platforms that make up our media infrastructures and the laws, policies, and economic and design practices that shape the platforms. Then, building on recent scholarship concerning the data practices of infrastructures used to predict, isolate, and channel attention, it details three characteristics of online communication about climate—noise, incivility, and ambivalence—and how each affects climate discourse.

Chapters 3 and 4 are about how climate activists are striving to overcome media and climate injustice and what we can do as a society to join those efforts. Chapter 3, "After Peak

Indifference," focuses on two movements, Fridays for Future and No Dakota Access Pipeline and how their fights for climate justice are also struggles for media and data justice. The chapter details the effort and skill that activists bring to their media engagement—building networks, creating content, garnering mainstream attention, expanding solidarity, and mobilizing—as well as the challenges and vulnerabilities they face on the uneven media playing field. Chapter 4, "Collective Imaginary," considers how to address the unequal power dynamics and to promote media justice so that we can more effectively and fairly engage climate-related issues. Revisiting the argument introduced here that we need to rethink our commitment to liberalism and its celebration of individual negative freedoms, chapter 4 underscores that the fight for climate justice is part of a long history of compelling challenges to the idea that priority and protections should be given to the freedom to pursue individual interests rather than our collective freedoms to pursue the public interest.

My aim is to make clear that in this crucial moment we must take stock of our options and rally all our resources to upend the status quo, which has become indefensible. In asking readers to consider the media and natural environments together, I hope to present both in a new light.

1

HOUSE ON FIRE

In the fall of 2019, President Donald Trump discussed climate change on *Good Morning Britain*. "I believe that there's a change in weather, and I think it changes both ways," he said. "Don't forget, it used to be called 'global warming.' That wasn't working. Then it was called 'climate change.' Now it's actually called 'extreme weather,' because with 'extreme weather,' you can't miss."[1]

Of course, carbon emissions and other greenhouse gases are doing all of the above: the past five years have been the hottest on record, and weather anomalies have become more common as temperatures continue to rise. These kinds of comments by a sitting U.S. president fuel ignorance among those who mistrust science and who take their cues from political leaders about who and what to believe.

Ignorance is often not simply the absence of knowledge; it can also be the intended consequence of cultural and political messaging wars. The tobacco industry famously engaged in the deliberate production of ignorance through advertising campaigns designed to manufacture doubt around the health effects of its products. Agnotology, the study of manufactured ignorance, examines how ignorance can be deployed as a strategic tool. It is most often used to obscure social problems that demand solutions that run counter to the interests of people in power—solutions that cut into bottom lines or campaign donation totals or election wins or all of these things. Obscuring and manipulating information have long been central features of the abuse of power, but the strategies honed by tobacco companies over the last half of the twentieth century are paradigmatic. They serve as a blueprint today for major campaigns, such as the one mounted since the 1980s against the science of climate change as well as those waged across various other economic sectors from the sugar industry to the firearms industry. A memo from tobacco company Brown & Williamson in 1969 read: "Doubt is our product, since it is the best means of competing with the 'body of fact' that exists in the mind of the general public[;] . . . it is also the means of establishing a controversy."[2] The strategy of spreading confusion to keep controversy alive and stretch out the era of booming bottom lines is kept alive today by industries using think tanks, lobby shops, and messaging gurus to successfully misinform the public and stall meaningful regulation.[3]

We tend to see the manufacturing of doubt as an aberration, a sly tactic of the corrupt, rather than what it is: an accepted set of professional skills honed over decades to confuse, obfuscate, and diminish the power of facts. Doubt as a product—the production of uncertainty where none exists—has deep historical roots, and it is still a centerpiece of today's communication landscape, characterized by a disappearance of shared standards for truth. It blurs lines between facts and so-called alternative facts or falsehoods, between knowledge and opinion, between truth and belief. And it feeds ignorance. When in doubt, people generally believe information that confirms rather than challenges their already-existing views[4] and allows people to avoid facts that challenge their way of life. Indeed, doubt is often the foundation upon which ignorance is manufactured.

Journalism as it is conceived in democratic societies ought to be an antidote to professionally induced ignorance. Whether the central role of journalism is to spark conversation and social cohesion, serve as a watchdog or fourth estate, or be a mirror to events, journalists are meant to assemble the relevant facts about issues and events by following certain procedures and rules. In the simplest conceptions, journalists fairly present opposing sides of the story, gather information from bureaucratically credible sources, and remain objective by separating the reporting of facts from opinions and value judgments when conveying the story to the public.[5] Indeed, for a large stretch of the modern era, a main product offered by journalism publishers has been a commitment to reliably accurate information. It thus makes sense that as we have come to recognize the extent to which falsehoods have shaped the discourse on climate change, many have blamed journalists for failing the public and even for being complicit in the spread of lies through either incompetence or gullibility or myopic dedication to norms and procedures.

Journalism has gone through extraordinary transformations over the past three decades as the mass-media news environment has given way to the networked news environment—transformations characterized by a widely expanded range of content producers and formats, shifting practices, shrinking newsroom budgets, and booming growth in tech platforms' power over news content and distribution. During these same decades, climate change transformed into a fast-moving crisis, and the story became a prominent part of the news cycle. While journalists struggled to keep up, they also played a key part in polluting the information landscape by at times amplifying lies meant to manufacture doubt.[6] After decades of flawed coverage and in response to a changing information landscape, though, journalists are now rethinking how best to cover the crisis, to adapt to the changing environment by shifting their practices to respond to the changing media environment, and to raise the quantity and quality of coverage. In the vast, complicated, and often toxic mediated space that journalists inhabit, these shifts seem something like sweeping the porch while flames roar through the rafters. Considering the larger media landscape in which journalism operates, it's worth asking: Are journalists really up to the task of addressing the contemporary information crisis? Can we expect journalism alone to put out the fire?

This chapter is about changes in climate journalism over roughly the past decade, especially but not exclusively among journalists in the United States. There is no way to tell that story, though, without first telling the story of how journalism failed and became a tool of climate-change denial amplification, which led to the current moment of reckoning. After unpacking the mechanisms of culturally produced ignorance around climate change and journalism's sometimes complicit role, this chapter details journalistic attempts to collectively respond to the climate crisis. These responses and recent shifts in climate reporting underscore both the possibilities and limits of journalism in the context of the political and material conditions in which journalism today operates—the corporations, tools, platforms, policies, and economics that shape journalist practices, products, and impacts.

JOURNALISM ENLISTED IN THE WORK OF MISINFORMATION

Beginning in the 1970s when Exxon scientists discovered for themselves that carbon emissions were warming the planet, the company spent millions of dollars on disinformation campaigns aimed at muddying the findings by climate scientists.[7] In the five decades that followed, as carbon emissions and temperatures climbed, other fossil-fuel companies joined the charge. In response to their efforts, sophisticated counterefforts were deployed to promote science-based climate information and action, including six rounds of IPCC reports in which scientists across fields and around the globe collaboratively assessed the state of scientific, technical, and socioeconomic knowledge on climate change; annual UN climate summits (or COPs) where world leaders gathered to hash out climate agreements; global activist movements; and countless others. Even so, spending on disinformation campaigns by the fossil-fuel industry reached its pinnacle in the three years after the Paris Agreement of 2015, which set out a global framework to avoid climate change by limiting global warming. The five largest publicly traded oil and gas companies (ExxonMobil, Royal Dutch Shell, Chevron, BP, and Total) invested more than $1 billion between the signing of the Paris Agreement and the end of 2018 on a sprawling network of front groups and on so-called narrative capture and lobbying, including reams of tendentious research.[8] The industry's hallmark misinformation tactic was and still is to use conservative think tanks such as the Cato Institute, the Heritage Foundation, and the Heartland Institute to pay willing scientists to produce or sign off on bunk research that can be used to marshal against any new emissions regulation. The research is pushed by front groups that target U.S. Capitol Hill and state legislatures and go door to door to bring around voters who lend a grassroots sheen to the efforts.[9]

A host of structural aspects of the current information ecosystem makes it vulnerable to information pollution. Private companies that own and operate our news and information infrastructures and outlets (including both legacy media organizations and, more recently, tech platforms) bear no formal responsibility to serve the public interest but are responsible to shareholders to make profits. This means they cater to audiences whose attention attracts ad revenue or whose data are deemed most valuable, leaving vast local news deserts ripe for exploitation.[10] For example, in August 2022 Chevron launched the website Permian Proud with content aimed at an oil-rich but news-poor area of Texas. The site mimics the format and style of a local news site, featuring feel-good stories about local events and encouraging readers to submit information on community events and initiatives. It also features heavy doses of industry propaganda and greenwashing—stories that exaggerate the scope of the company's solar-power and water-recycling projects and that falsely paint a promising picture of the future for the oil and gas industry in the region. Permian Proud is run by the San Francisco–based public-relations firm Singer Associates, which since 2014 has also churned out copy for another Chevron-funded faux-news site in Richmond, California, the location of the company's 2,900-acre petroleum refinery. This practice of using paid media to target crucial geographic oil-producing areas is not uncommon, and such sites often pop up in areas bereft of any sources of news, thus exploiting the well-documented fact that lack of local news outlets makes people more vulnerable to disinformation because they are pushed to rely on less trustworthy sources of information.[11] One-third of Texas newspapers have closed over the past two decades,[12] and twenty-seven counties in Texas have no news outlets.[13] Roughly half of these counties are in the Permian Basin, where Permian Proud's target audience resides. In addition to Permian Proud, in 2022 Chevron expanded its paid-media strategy by launching several media-based initiatives, including a paid partnership with Houston Public Media to generate "news" stories on how the energy sector is supposedly working toward a lower-carbon future, filled with oil company greenwashing techniques and story lines. Houston Public Media later suspended the series, citing criticism and admitting that airing paid content in the form of news "created confusion."[14] Chevron-funded propaganda masked as news not only fools people but also makes it harder for actual journalism to keep its competitive edge. "Previously, news has been trusted (generally, at least) to provide some semblance of facts on what is happening in the world," write the media scholars Stephanie Craft and Morten Kristensen, "but as journalist-produced news loses its privileged status, it becomes merely one type of content competing for attention with many others, its distinctiveness lost in a cacophony of media noise. The result is a de-centered media landscape in which all media content is roughly equal."[15]

Propaganda and misinformation funded by the fossil-fuel industry are not the only pollutants to muddy the track and obscure the causes and effects of the climate crisis. Right-wing political parties and movements that support nationalist agendas cultivate distain for and mistrust in journalism among their supporters.[16] Journalists face harassment and violence in the United States and around the world,[17] and polls show a declining level of trust in journalists among members of the public.[18] Right-wing political leaders also generate doubt in science by openly dismissing scientific findings and recommendations. When elected in 2019, President Jair Bolsonaro of Brazil dismantled several government divisions dedicated to climate change and named cabinet members who are openly hostile to the fight against global warming. Minister of Foreign Affairs Ernesto Araújo claimed that global warming is left-wing "dogma" used to "suffocate the economic growth of capitalist, democratic countries."[19] Far-right leader Thierry Baudet in the Netherlands frequently rails against "climate change hysteria." And, of course, in his role as president Donald Trump once claimed that climate change is a "hoax" and asserted that "global warming was created by and for the Chinese in order to make U.S. manufacturing non-competitive."[20] The proliferation of these doubt-sowing political forces presents politics to publics through a lens that jumbles fact and fiction and fosters the sort of distortion and obfuscation that runs rampant in today's information landscape. The fact that there are still debates about whether and to what extent the climate crisis exists is exactly what the misinformation industry has been working toward the whole time and achieved with its no-holds-barred messaging efforts.[21]

Given all of this, the manipulation of journalists is no surprise. Indeed, journalism as it has been practiced and professionalized in the United States is particularly vulnerable to specific misinformation strategies. The appearance of objectivity, the linchpin of U.S. journalism, is meant to be achieved through a combination of bureaucratically credible sources and a commitment to fairness through balance. But the culture of objectivity ideally demands fairness and balance only within the limits of consensus or the "common sense" of a particular political system.[22] Contrary to this ideal, combined commitment to authoritative sources, many of whom have a vested interest in the preservation of the fossil-fuel economy, and to the demands of the market-driven news industry created conditions for U.S. journalists to fall prey to political and corporate obfuscation and denialism. As Max Boykoff and Jules Boykoff demonstrate in their study of global warming in the elite U.S. press from 1988 to 2002, the adherence to balance led to biased coverage of both anthropogenic contributions to global warming and the need for action. "This bias, hidden behind the veil of journalistic balance," they conclude, "creates both discursive and real political space for the U.S. government to shirk responsibility and delay action regarding global warming."[23]

The way climate change has been covered in the United States since 1988, when the climatologist James Hansen made it a public issue by testifying to its existence in front of the U.S. Senate, underscores a vulnerability tied to two assumptions at the base of the country's journalism profession: that policy debate is an essential part of the democratic process and that journalism for this reason should be practiced in a way that propels policy debate. In the case of climate change—as a subject of policy debate and as a policy debate worthy of news coverage—the result of these assumptions has been absurd and irresponsible. The Harvard geologist and science historian Naomi Oreskes compared the debate to an argument over the final score of a baseball game. "If the Yankees beat the Red Sox 6 to 2, journalists would report that," she wrote. "They would not feel compelled to find someone to say, actually, the Red Sox won, or that the score was 6 to 4."[24] I would add that journalists would also not feel compelled to report in a straight way that outside the stadium a man in a suit was presenting a white paper produced by a think tank, arguing that it's not possible yet to report a winner of the game because data are still being collected and the score keepers present at the game may or may not be reliable.

In the case of climate change, the problem is not just that a debate was manufactured where none exists. The more daunting problem is that for a democratic system to function, so the thinking goes, the appearance of debate is necessary. The public-relations scholars Melissa Aronczyk and Maria Espinoza write that, "in this context, the task of PR is to mirror democratic structures of advocacy, by representing various viewpoints, providing information, and soliciting opinion. That this information and these viewpoints are plainly unscientific, that these truths are clearly 'inconvenient,' appears less important than adhering to the values of democracy."[25] It is the link between the appearance of debate and our ideas about how democracy should look that explains in part the long string of "both sides" climate stories that have dominated news coverage for decades and why climate misinformation and confusion and skepticism have won an inflated level of visibility in the mainstream mediasphere despite the clarity and near consensus on the facts reported in legitimate scientific literature over the same period. The appearance of debate also benefits from the fact that it fits well with the mainstream press's market logic. Journalists fell into the trap of amplifying bogus information not because the lies were so convincing or compelling but rather because debate sells and the opinions of the powerful must be heard.

Maintaining the appearance of debate in climate journalism is driven by a commitment to the information-deficit model: the notion that more and better information will fuel more effective climate policy and action. But through the lens of the information-deficit model, public uncertainty and skepticism toward science, including environmental science, look like a lack of sufficient knowledge about science. Two assumptions are the result: the marketplace of ideas will ensure that the highest-quality information will rise to the top and gain widespread acceptance, and the person who fails to properly recycle or opposes renewable energy or bold energy-efficient upgrades simply has not been properly informed on the related issues and consequences. And if these assumptions are true, then the best action to take is to provide more or better information in a more accessible way.[26] In the context of contentious science, policy, and politics around the climate crisis in the United States, commitment to the logic of the information-deficit model creates a space for skeptics and denialists because it shifts the story from climate-related facts, where denialists have no place, to climate-related debates, where they are the drivers of the conversation.[27]

Constant flows of information and the commitment to debate in turn invite anyone—including those spreading disinformation—to contribute to the "debate" and the bias toward those who claim to speak authoritatively: these foundational practices and assumptions create and maintain the conditions for the manufacturing of doubt, which turns journalists into inadvertent collaborators or signal boosters.[28] As Whitney Phillips found in her seminal study of reporting on online manipulators, "The choices reporters and editors make about what to cover and how to cover it play a key part in regulating the amount of oxygen supplied to the falsehoods, antagonisms, and manipulations that threaten to overrun the contemporary media ecosystem—and, simultaneously, threaten to undermine democratic discourse more broadly."[29] That is, when journalists are not on guard against how their practices and values are manipulated to enlist them in the spread of polluting information, they become a central part of the problem.

All of this leads to why scientists, environmentalists, and academics routinely criticize journalists for failing to properly cover climate change. Indeed, it has come to the point, three decades into the issue, where journalists on some level have conceded the point and become perhaps the profession's most vociferous critics. As Mark Hertsgaard, the executive director of the climate journalism partnership Covering Climate Now, puts it, "Better news coverage is an essential climate solution, a catalyst that makes progress on every part of the problem—from politics to business, art to activism, and lifestyle change to systems change—more likely."[30]

Activists are also calling attention to the problem. Police arrested roughly seventy Extinction Rebellion activists in Manhattan during a protest outside the *New York Times* building in Times Square in June 2019. The group published a "Standards for Media" guidebook that implored journalists to produce more coverage of environmental issues and more solutions-based coverage: "Dear print media, it's time to take a stand. Declare a climate emergency and report accordingly."[31]

The assumption underlying this finger pointing is that public ignorance and apathy are evidence of journalism's failure. *Why don't people understand? Why don't they care?* These questions have spurred important research and newsroom debate about what journalists' role should be in responding to the climate crisis and how best to produce coverage on it.

A STUBBORN PROFESSION EVOLVES

The widespread acknowledgment that the climate is in crisis has given journalists license to rethink their work.[32] Mainstream news outlets have demonstrated their ramped-up commitment to responding to the climate crisis in various ways. The *Washington Post*, the *New York Times*, and the Associated Press have established a desk dedicated to climate and the environment. *The Guardian*, in addition to its robust coverage of the climate crisis, has divested company holdings from fossil fuel, committed to achieving net-zero emissions by 2030, and most recently announced it would no longer accept ad money from oil or gas companies. Online outlets such as *Vice*, *Buzzfeed*, and the *Huffington Post* dedicate resources to covering the climate crisis and particular climate-justice issues, which resonate with their target audience, younger readers.[33] And new niche sites dedicated exclusively to covering climate-related issues are some of the most well-trafficked sources of climate information.[34]

According to data produced by the University of Colorado's Media and Climate Change Observatory, media coverage of climate change or global warming in newspapers, radio, and television around the globe reached its second-highest levels on record in October 2021 (behind December 2009), measured by the number of stories. And in the United States, coverage in all three media reached the highest levels of coverage on record to date.[35] An analysis of coverage over fifteen years (2005–2019) from seventeen sources in five countries (the United Kingdom, Australia, New Zealand, Canada, and the United States) found that 90 percent of media coverage accurately represented the scientific consensus that human activity is driving global warming. It also found that climate coverage was improving over time in terms of scientific accuracy, with coverage being significantly more accurate in 2010 than in 2005, with less space afforded to the perspective of climate-change deniers.[36]

This reduced focus on denialist perspectives is perhaps the most significant shift in coverage, but journalists are also changing their practices in other significant ways. They are increasing collaboration with one another and with scientists; they are developing new ways of relating to activism; and they are acknowledging that they are part of the process of meaning making rather than simply mirrors that reflect reality.

MOVING AWAY FROM DENIALISM

Several factors have triggered the move away from covering denialism and toward more substantial discussion of what can and should be done in response to the crisis. In the early 2000s, fossil-fuel companies began to change their marketing tactics. After decades of denial, they shifted to campaigns promoting the idea that individuals could solve the crisis by changing their own habits, another approach pioneered by the tobacco industry, which after years of denial blamed smokers for their addiction and its associated health impacts. Fossil-fuel giants such as Chevron, BP, and ExxonMobil have spent a fortune to convince the public that consumer choices and lifestyle changes will solve the problem. One infamous example is BP's "carbon footprint calculator," which blamed greenhouse-gas emissions on people's daily activities.[37] While the individual responsibility mindset took hold, the "debate" about climate change receded as a central feature of news reports with the absence of messaging from the industry that climate change doesn't exist. This fading of the debate as well as the increasing urgency of the crisis and audiences' decreasing tolerance of blatant falsehoods, even when they come from political and economic leaders, served to profoundly shift how the crisis is framed.[38]

Journalists today are more on guard against providing a space for points of view that distort the reality of the crisis. Journalists who cover the climate crisis tend to agree with the scientific consensus put forth by the IPCC, documented in its gold-standard reports, and with the panel's proposals for how best to address climate change.[39] As Michael Brüggemann points out, in these "post-normal" times journalists prize standards such as fairness and transparency over so-called objectivity and engagement over neutrality, with the aim of presenting a more accurate picture of the climate crisis.[40]

Newsrooms, for example, have formulated varying policies as they wrestle with climate denial. The *Los Angeles Times* and the *Washington Post*, among many other national and local papers in the United States, no longer publish letters or opinion pieces from deniers on their editorial pages. The BBC in 2018 circulated internal guidelines stating its official position: climate change exists, and reporters "don't need a denier to balance the debate."[41] Robert Eshelman, former *Vice* environment desk editor, explains *Vice*'s position: "If it's a piece about science, you will not see climate change deniers in it because there is no possible legitimate scientific point they could be making." *Vice* will cover the perspectives of politicians who deny the existence of climate change only as a form of watchdogging because, he says, "people need to know when their leaders are denying scientific reality."[42]

Public intellectuals working the climate-change beat outside the traditional journalism space have led the way in jettisoning objectivity norms. As Matthew Nisbet has pointed out, activist-authors such as Bill McKibben and Naomi Klein do reporting and writing that abandons attempts at neutrality. They position themselves as topic explainers, occasional champions of specific policy positions or causes, and interpreters and synthesizers of complex areas of research. They privilege expert and political logic over media logic, meaning that they do not play by the historically received rules of the journalism profession and are not afraid to effectively participate in political debate.[43] They can take this approach in part because they focus their work on analysis and synthesis, making themselves less dependent on sources and freer to challenge the status quo.[44]

COLLABORATION

Over the past two decades, journalism outlets have increasingly experimented in collaboration, shifting away from the professional competition that has long shaped the market-driven news industries. Collaborative news production has risen in part as a way to make the most of limited resources and to extend the reach and impact of reporting in the digital era, during which the communication-information marketplace has exploded with options and journalism business models have fallen apart. Some of the most high-profile examples of collaboration, such as the money-laundering and tax-evasion stories tied to the Panama and Pandora Papers leaks and examined by the International Consortium of Investigative Journalism, leverage collaboration to meet the reporting challenges posed by massive data sets and the need to work across borders on regional and international issues.[45] Collaboration was often a key feature of public journalism projects in the 1990s, in which certain stories featuring elections or pressing community concerns were seen to be enough of a matter of public interest that competitors were justified, even morally compelled, to collaborate.[46] This attitude is echoed in the way climate journalists today are collaborating in the effort to improve the quality and impact of coverage. In Florida, a state intensely experiencing the impact of climate change, six major news outlets with different owners announced in 2019 that they were pooling their resources and sharing their reporting on how climate change will affect the state's agriculture sector as well as its coastal towns and cities. Within six months, the number of outlets involved had tripled. That same year, *Columbia Journalism Review* and *The Nation* magazine announced Covering Climate Now, a partnership that as of 2023 included more than five hundred outlets. The project's aim is to focus more media coverage on the climate crisis.[47] Also in 2022, the one-hundred-year anniversary of the Colorado River Compact, a management agreement among the seven U.S. states that the Colorado River runs through, the Associated Press partnered with media outlets in the U.S. West to publish a series of stories on the drought-stricken river, addressing what is at stake for the more than 40 million people who rely on the river for water and electricity.[48] These types of newsroom cooperation and collaboration related to the climate crisis stretch back at least to 2009, when fifty-six newspapers from around the world joined together to implore political leaders to reach a meaningful agreement at that year's UN Climate Change Conference (COP) in Copenhagen. The editorial signed by the outlets encouraged politicians to take action: "[You] have the power to shape history's judgment on this generation," it read. "[We are either a generation] that saw a challenge and rose to it, or one so stupid that we saw calamity coming but did nothing to avert it. We implore [you] to make the right choice."[49]

Climate journalists tend to be invested in improving coverage everywhere, not just at their own outlets. *Inside Climate News*, for example, is committed to a public-service model, making its content available for free. Yale Climate Connections creates content with the aim of spreading it far and wide at no cost to the outlets that run it. In 2014, it transitioned from a web-only site to a radio program that runs each weekday and tells ninety-second stories after it found that many radio stations across the country have holes of that size to fill, where they can either provide local news updates or slip in a program like the Climate Connections–produced stories. While some collaboration efforts focus on specific projects, such as the Florida collaborations, more often they are directed toward climate coverage more generally, whereby outlets offer services and share content with one another.

There is also an increase in collaboration between journalists and scientists. Journalists are more committed to conveying science accurately, and scientists are more willing to engage with media and the public. Some scientists have gone through a reckoning like that of journalists, wherein they have had to construct new professional practices and strategies around the need to counter misinformation. Leah Ceccarelli calls scientists engaged in such efforts *scientist citizens,* describing them as "experts who recognize a responsibility to act for the public good by stepping out of the technical sphere to make arguments about science-related matters in the public sphere."[50] These scientist citizens have forged collaborative alliances across academic disciplines as well as with journalists.[51] Leo Hickman of the climate news website Carbon Brief explained, "There is a more matured and nuanced understanding amongst scientists around how the media works and almost how to play the media game."[52] Indeed, more scientists are using social media to engage directly with journalists and explain their work, to correct inaccurate coverage, and to provide source material for climate stories.

Mat Hope, former editor of the climate-focused online news outlet *DeSmog UK,* says these improvements are in large part due to scientists having more positive interactions with journalists. "I think it started out by [scientists] getting confidence because they're being contacted by journalists who are specialists. So [the journalists] would ask fair questions essentially. [The scientists] wouldn't have to start every conversation with explaining what climate change was."[53] Thus, a positive feedback loop is set in motion whereby improved coverage fuels a more hospitable environment for scientists, which in turn further improves coverage.

NEW WAYS OF RELATING TO ACTIVISM

Related but distinct from increased collaboration with one another and with scientists, some climate journalists are developing less adversarial, more productive relationship with activists.

Earlier journalism research on the dynamic between activists and mainstream journalists documented the prevalence of a "protest paradigm," emphasizing that journalists tended to draw attention away from the actual issues at hand and instead framed protests as deviant or frivolous behavior.[54] This approach simultaneously delegitimized the activists and legitimized the elite news sources' points of view and agenda. Climate journalists today work at having a reciprocal relationship with activists, giving them information to support their cause and in turn getting from them a boost in attention to content and issues. Leo Hickman of Carbon Brief, for instance, points to the "Greta Effect," a label referring to the popular climate activist Greta Thunberg, whose ability to bring attention to climate science and rally youth political engagement has given her enormous power in the media landscape. In 2020, she tweeted a link to a Carbon Brief story, which then quickly received more than 8 million hits. Hickman describes how Carbon Brief also supports Thunberg, who, he says, has come to trust Carbon Brief content enough to use it to back up her own positions. "That's been quite exciting. We aim to try and do more of that if we can."[55]

This is not to say climate journalists routinely disavow the norms of professionalism followed by their colleagues on other beats. Vernon Loeb, executive editor of *Inside Climate News*, describes what he and his colleagues do: "We are not activists. We are journalists—non-partisan non-opinion reporting journalists." But he does see their coverage as having various types of impact: "There's direct impact which you write about somebody and they stop an activity or they embark upon an activity. There's indirect impact where somebody posts what you've written or mentions what you've written and then there's the impact on readers."[56]

MEANING MAKING

Journalists are also embracing their role in shaping attitudes toward the environment and climate change. In May 2019, *The Guardian* announced that it was updating its style guide. Editor in chief Katharine Viner explained that the idea was to be more "scientifically precise" and to communicate more "clearly" with readers. The paper now uses the terms *climate crisis* and *climate emergency* instead of *climate change*; *global heating* instead of *global warming*; *wildlife* instead of *biodiversity*. *The Guardian* has also added global carbon-dioxide-level data to its daily weather pages. In a post announcing the changes, the paper cited the input of scientists and the UN and added that the activist Greta Thunberg helped inspire the change.[57]

Media studies research on the politics of representation underlines the power that language, ideology, and professional norms have in creating meaning from issues and events.[58] This added attention to the symbolic power of language is increasingly evident in coverage of climate politics. Journalists delivered high-profile reporting, for example, on the fact that the Trump administration removed the term *climate change* from federal government websites and the word *science* from the Environmental Protection Agency mission statement and its reports.[59] And, of course, Trump's use of language to downplay the urgency of the crisis was also in full display in his *Good Morning Britain* appearance in 2019, described in the opening of this chapter. When he quipped about shifts in terms from *global warming* to *climate change* to *extreme weather*, he was attempting to project doubt by claiming language inconsistencies and suggesting that the word *extreme* can only in this context be an exaggeration. Language matters for anyone attempting to shape how we understand the world.

Beyond specific word choices and combinations, climate journalists are more often and more readily making connections between racial and economic inequity and the effects of the climate crisis, a shift due in part to the strengthening of activist movements and messages as well as the growing awareness of the connections between climate justice and racial justice. As the environmental reporter Rachel Ramirez puts it, "Environmental justice reporting demonstrates that the issue of environmental racism impinges on every story—whether it's race, housing, economics, healthcare, or immigration."[60] With this growing awareness that experiences of the climate crisis are always entangled with other issues, journalists such as Ramirez are thinking beyond traditional beats to consider climate economies, climate justice, and climate politics, to name a few prominent intersections. In doing so, they are laying the groundwork for new ways of telling climate-change stories.

Laurie Goering of the Thompson Reuters Foundation says a large part of the foundation's climate coverage ends up being about climate justice because it focuses on legal battles around the social and human rights implications of the climate crisis. "We almost don't see it sometimes because that's always been at the core of what we've been writing about," she says. "But you're starting to see these two groups link up." In addition to being a financial and legal story, she explains, the climate crisis is ultimately a story about social equity. "Like we've seen in the United States, (the) nexus between environmental justice and racial justice is a central part of the coverage."[61]

Vernon Loeb links *Inside Climate News*'s change in coverage of justice-related issues directly with the rise of Black Lives Matter protests and the ensuing rise in awareness of issues of structural racism and injustice: "This moment we're in now has helped us see that environmental justice is something we ought to make an ongoing hallmark of our coverage along with say science. It's such an undeniable part of the climate story that people who are disproportionally affected are people of colour."[62]

Coverage that underscores the links between the climate crisis and financial, social, political, or legal inequity is part of a wider shift among some news media to acknowledge systemic inequities that mainstream media previously ignored. For example, in coverage of police violence against Black people in the United States, news outlets now more frequently frame the many uses of deadly force and police racism as systematic, the result of a broken system rather than of individual police officers and their victims.[63] This shift in emphasis has occurred at least in part because, as Allissa Richardson describes it, "news is no longer solely the domain of the elite. African Americans have seized social media platforms to elevate their discourse to a national level, and beyond."[64] The connection between racial injustice and climate-change impacts are part of that change in discourse.

This increased attention to how the crisis is represented—the language uses, frames deployed, sources interviewed, attention spent—suggests journalists are more readily including social and political implications, not just scientific expertise, to shape understandings of the climate crisis and its impacts. It also points to an evolution in coverage and in journalists' commitment to getting the story right more than to appearing balanced or objective. Climate journalism is experiencing in fits and starts a move toward what I have described elsewhere as activism on behalf of the evidence- and experience-based facts instead of a particular cause or perspective.[65] Journalists work toward producing stories that present reality from a perspective, anchoring them to evidence and experience rather than echoing the points of view of authoritative sources. Consider the ongoing COVID-19 pandemic-related urgent science and politics reporting that refuses to fall into the "both-siderist" balance trap and thus coaxes reluctant politicians, propaganda networks, and sectors of the public to accept the scientific realities around the pandemic and take dramatic public and private action to meet the threat. Climate journalism is moving toward similarly engaging with the idea that we have entered the era of climate crisis and that journalists have a responsibility to treat the climate story as a scientifically based threat to public health and safety and to economic and political stability that is every bit as real as a pandemic.

DOING JOURNALISM IN THE DIGITAL WILDS

The evolution of climate journalism—away from denialism and toward greater collaboration, more productive ways of relating to activism, and more reflection around meaning making—demonstrate the extent to which the climate crisis is shaping professional judgment and norms. This is happening in important and productive ways despite constraints that include traditional journalism practices and values, media- and tech-industry profit imperatives and political alliances, and the power wielded by entities dedicated to manufacturing doubt and ignorance. These emergent practices underscore a flexibility in the journalism profession that enables members to pull together and mobilize resources, align thinking, and alter their approach to the work.

But even the best journalism must operate in the pollution created by well-funded and highly organized forces working not just against climate crisis solutions but also more broadly against a healthy information landscape. The powerful, mostly organized forces of climate denialism are an early and spectacularly successful example of how issues are negotiated in today's mediated landscape. "Seemingly overnight, norms of reasoned debate between competing viewpoints have given way to willful distortion and lying," write the politics and media scholars W. Lance Bennett and Steven Livingston. "Institutional arenas designed to express and resolve political differences are disrupted and fail to provide the gatekeeping roles that once kept discourses bounded by a more or less shared set of institutional norms and processes."[66]

But, in fact, the disruption of norms of debate did not happen overnight. Our current epistemic rift—which is affecting common notions of what we value, whom we trust, how we know things, and what we believe—has been taking shape for decades. The campaign against climate action is just one example in a decades-long effort to undermine trust not only in science and journalism but in all liberal democratic institutions, including not least of all governments and courts. Traditional seats of expertise and authority that set standards for how knowledge is produced—such as peer review in science, procedures of evidence in the courts, norms of fairness, and verifiability in journalism—have come under attack precisely because of their role as arbiters of information.[67] Some of this erosion of trust is rooted in institutional failings. Well-documented government scandals of lies and coverups have stacked up since the Vietnam War era—the Pentagon Papers, Watergate, Iran-Contra, the deceit and corruption associated with the war on terror, and the massive expansion of unconstitutional surveillance after September 11, 2001, are on a very long list. Journalism played a large role in exposing the abuse of power at the root of those scandals, but the scandals also in many cases at least implicated journalism in part by exposing the weaknesses and unreliability that comes of its role as authority's messenger. Investigative triumphs and exposures have also paradoxically contributed to the wariness and suspicion toward government power and the deepening skepticism toward authority of all kinds that blossomed in the United States in the post–Vietnam War era. The forces working to obscure climate science and gum up the machinery that would address climate change feed on this zeitgeist.

The assumption that more and better information can overcome public skepticism toward established facts and science ignores the larger context in which journalism and other forms of media operate today, including the ongoing and concerted efforts to manufacture ignorance and doubt as well as the ways that widespread reliance on corporate communications platforms undermine independent journalism. The idea that good information will triumph is baked into journalism and thus not easy to see past. Indeed, what kind of journalism would fail to prioritize the work of delivering quality information or fail to foster the expectation that such work would win out in the end over the dissemination of bad information?

We have, however, seen time and again that more and better information will not change our climate trajectory. People make sense of facts differently. Knowledge does not necessarily translate to power, especially when ignorance is intentionally being manufactured. Journalism may play a less central role in creating ignorance today than it did over the past several decades because, among other reasons, it more accurately represents science and refuses to host debates about whether global warming exists and is caused by human activity, which it sees as indisputable facts. But that doesn't mean journalism is poised to be part of the solution to the crisis. That is a tall order for journalism, especially as it is currently conceived.

Efforts to manufacture ignorance around climate change is not an aberration but a long-term strategy that has shifted and adapted to the changing political and information environment, and it is not likely to go away anytime soon. Now, in addition to spreading through the pages of the newspaper, doubt and ignorance also spread through social media feeds; press conferences featuring faux scientists will continue, as will the spread of amateur videos touting antifacts and climate conspiracies. Offline harassment and assault of scientists are joined by steady flows of online attacks by bots and trolls. Outright denial has been replaced mostly by massive greenwashing meant to cow publics into believing individuals bear responsibility for mitigating the crisis and that the fossil-fuel industry is the leader rather than the enemy of climate-crisis action.[68]

And, of course, journalism's increasing reliance on networked technology platforms to host and deliver content also constitutes a major shift in the realities of journalism.[69] As the law professor and technology and society analyst Frank Pasquale argues, that relationship will continue to drain power from journalism. "A general trend toward media revenue decline (and platform revenue growth) makes a new endgame apparent," he writes. "Online intermediaries [will become] digital bottle-necks or chokepoints, with ever more power over the type and quality of news and non-news media that reach[] individuals."[70] Indeed, social media platforms create attention economies that grant disproportionate amounts of power to a narrower range of actors. In vying for attention in the rush of change, journalism is reinventing itself. In some cases, this reinvention is in ways meant to improve the quality of coverage, but it often is meant to optimize popularity on social media platforms, which doesn't seem like a plan that will deliver long-term business growth and increased influence or one that will best serve the public interest. In a kind of Faustian bargain that legacy media have made with social media, they work toward getting clicks and exposure via social media, while they feed into a system that subverts the information ecology in ways that make it more difficult for good information to rise to the top.

Indeed, this trend toward media-revenue decline and platform-revenue growth points to the role digital platforms play in the steady undermining of democratic politics. The rise of digital technologies initially prompted scholars concerned with public life to celebrate the way they imagined these tools would strengthen democratic communication, as in the rise of niche climate news sites. Many of the same optimistic observers now write, however, about how networked infrastructure—its data practices, platforms, and algorithms—poses new-level challenges to the larger project of democratic society.[71] As Zeynep Tufekci wrote in a *New York Times* editorial in 2018, "The real problem is that billions of dollars are being made at the expense of the health of our public sphere and our politics, and crucial decisions are being made unilaterally, and without recourse or accountability."[72]

Let us now return to the question posed at the beginning of the chapter: Should we expect journalists or journalism to win the battle against the information wildfire we're living through? The evidence piled up over the past three decades strongly suggests we should not. Expecting climate journalism, no matter how much improved, to solve the climate-information crisis ascribes a mass-media solution in a networked environment. It ignores the context in which contemporary journalism operates, which includes both challenges left over from the mass-media era and new ones present in the social and technical infrastructures where journalism today exists. Chapter 2 explores the many barriers that prevent good information from "rising to the top" of the information space and argues that the faith in this marketplace of ideas has never been more misplaced.

2

NOISE, INCIVILITY, AND AMBIVALENCE

In July 2020, the CEOs of Twitter, Apple, Google, and Facebook testified to the Judiciary Committee of the U.S. House of Representatives in defense of their business practices, claiming that they did not amount to anticompetitive monopolies.[1] "This is the moment," stated the *New York Times* technology and regulatory policy reporter Cecilia Kang, "when . . . the captains of the biggest companies in technology, just like we saw the heads, the captains of the biggest companies of the tobacco industry, have to come before Congress—and really defend themselves as companies that are potentially harmful to society."[2] She was referring to the House hearing in 1994 at which executives of the seven largest U.S. tobacco companies infamously stood up and testified falsely that cigarettes made with nicotine as a prime ingredient were not addictive.[3]

There are clear instances now of the four tech companies buying up the competition to make themselves stronger and discriminating against rivals on their own platforms. Typically, antitrust concerns have to do with lack of competition driving up the cost of the products and lack of competition driving quality down. But because the "products" produced by tech platforms don't cost publics in terms of money but in rather personal data, a different and complex set of issues has come into play. There is evidence, for example, that the rise of big-tech monopoly power has weakened workers' bargaining power and slowed rates of innovation.[4] There are also concerns that big tech has hobbled new industry's ability to turn enough profit to stay afloat. That same year, three of the companies called to testify—Apple, Google, and Facebook—were ranked the world's largest news and information companies based on revenue. A decade earlier, Rupert Murdoch's News Corporation was the largest news media company in the English-speaking world, but over the ensuing ten years the digital ad revenues generated by Google and Facebook as well as the service revenue generated by Apple's app store, iCloud, Apple TV, and Apple News, among other platforms, vastly outdid the Murdoch company's or any other news company's revenue.[5] Thus, the hearing was not only about the possibly illegal practices that led to the massive growth and monopolistic control of the online environment but also about how today's digital platforms serve as the infrastructures on which our political, commercial, professional, and social interactions take place and about the impact these companies have on our communication environment and on journalism in particular.

In chapter 1, we saw how journalists strive to improve coverage, adapting their practices in response to shortcomings of the past, spurred by a faith that delivering more and better information will lead to rational deliberations based on facts. This chapter examines what they overlook in that way of thinking—the social and material contexts that affect both journalism and publics. In particular, it explores the way platform infrastructures, combined with long-brewing epistemic rifts regarding what constitutes truth and what sorts of sources are trustworthy, affect journalism and make some types of publics more possible than others.

Infrastructures in the digital era are drastically different than those that provided the scaffolding for life in the predigital era. Where infrastructures were once relatively stable universal services delivered by monopolistic companies, today they are "platformized," meaning that they are made up of web-based architectures that facilitate connection, programmability, and data exchange. So although big-tech giants such as Google and Meta may be monopolistic in the sense that have exploited the power and reach of platforms to become the contemporary equivalents of the railroad, telephone, and electronic utility monopolies of the nineteenth and twentieth centuries,[6] the actual platform infrastructure they provide comprises constantly shifting interaction and therefore is much less stable and much more visible. We can more readily see and feel their influence when, for example, Meta/Facebook changes the design of its newsfeed, or Twitter bans high-profile troublemakers, or Google changes its web-page-ranking algorithm.

In the same decades that saw this shift to platform infrastructures, common notions of what we value, whom we trust, how we know things, and what we believe have receded. As elaborated in chapter 1, we are experiencing an epistemic rift. Today facts are less influential in shaping public opinion than are emotional appeals, and accuracy and transparency have been replaced by algorithmic sorting that delivers people what they want to hear. Climate-change denial, for example, takes overwhelming scientific evidence as just another opinion.

To be clear, the epistemic rift that characterizes contemporary publics was not caused by platformized infrastructures but rather by a combination of technological and social forces—the content we are exposed to and engage with online; the stories journalists choose to tell and the incentive they have to tell them; the algorithms that promote certain types of content over others; the regulations that govern speech, media outlets, and political communication. How well or how poorly we govern ourselves, share concerns, encourage or sanction behavior is tied to how well our communication systems work. Mike Ananny writes, "Today, these systems of communication—these systems of self-governance that make publics—increasingly live within privately controlled infrastructures. These infrastructures create the conditions under which people make meaning. They make some publics more likely than others."[7] The tension between privately controlled platform infrastructures designed to maximize profit and the publics that depend on these companies to act in the public interest is perhaps so obvious that it is often ignored.

The House Judiciary Committee hearing in 2020 was a start, but, given the prominent role tech companies play in shaping publics, we need to take a deeper critical perspective that includes but goes beyond questions of antitrust violation. We also need to get beyond the notion that good-information journalism will organically rise to the top in the marketplace of ideas. This chapter first provides background on the historical relationship between publics and media infrastructures and on the changing dynamics of these relations. Then, building on recent scholarship concerning the data practices of platformized infrastructures used to predict, isolate, and channel attention, it details three characteristics of current online communication about the climate—noise, incivility, and ambivalence—and how each shapes online climate discourse.

SHIFTING SOCIOMATERIALITY OF PUBLICS

The Public Sphere

The public sphere, the theoretical communication space located between the state and private spheres that Jürgen Habermas developed in his book *The Structural Transformation of the Public Sphere* (published first in German in 1962), is one of the most enduring and debated concepts associated with democratic theory. It gives us a normative ideal through which to consider the circumstances wherein something akin to a healthy exchange around issues of common concern can exist. To Habermas, the public sphere is the imagined realm in which "the public organizes itself as the bearer of public opinion."[8] His version of the public sphere is often criticized for exaggerating the extent to which an actual public sphere ever existed in its ideal form, emphasizing the idea of one overarching (political) public sphere, and ignoring the existence of other public spheres (literary, working class, Black, feminist, etc.) that shape political life.[9] It also overplays the role of rationality as the dominant arbitrator of public issues.[10] In recent years, Habermas has updated his thinking in response to critiques and to the changing media landscape, and reams of scholars have built on his original concept to develop a more robust understanding of contemporary public spheres and the role of media in their formation, maintenance, and interaction with one another.[11]

What the concept of a public sphere does offer, first, is a way to talk about the actual and imagined spaces where people come together to act as publics, engage with issues of common concern, and form opinions. Second, despite its historical inadequacies, it urges us to consider how this space and the material and social conditions that define it do in fact shape rather than simply facilitate what goes on in such a space.

Before Habermas first introduced the concept of the public sphere, Hannah Arendt offered her own version of public coming together in a creative and culture-forming space, "where I appear to others and others appear to me" and where a polis is built not around a common physical location but rather around people acting and speaking together in whatever space they may be.[12] For her, the public realm is made of what she called the "stories"—artifacts, relationships, and culture—that people create and hold in common.

The institutional, cultural, economic, social, and technological regimes put in place by mass privatization and commercialization in the twentieth century took a heavy toll on the public sphere. With the rise of mass media, access to the means of producing and distributing content—printing presses and airwaves—became very expensive and thus reserved for those with wealth or political power. Costly production led to the formation of a professionalized class of journalists and elite owners of media. The one-way flow from producer to audience provided limited opportunity to talk back. Publics appeared to one another through the lens of media professionals as mass media eclipsed forms of local news that had previously been the norm—news that came in the form of pamphleteering, letter writing, local newsletters, and salon-style or public-square-style conversation. A global system of technologies was accepted by the vast majority of people around the world and produced styles of content and content delivery that, in hindsight, alienated much of the public from active media engagement even as it was designed to ensure high-quality information and the privileging of authority meant to establish and maintain the public trust.

As noted earlier, critics have documented the way this reliance on authority has privileged the voices and perspectives of the political and economic powerful—mostly white men—at the expense of everyone else. In the case of climate communication, this privileging resulted in highlighting the disinformation generated by fossil-fuel industry think tanks over the voices of those on the frontlines of the impending climate crisis. It also meant that very little attention was paid to the lived experience of the communities facing the impacts of environmental degradation. This reality endures, particularly in the realm of broadcast news, which caters to older audiences and tends to feature sources that are considered bureaucratically credible (i.e., are in positions of power). A study by the nonprofit watchdog group Media Matters found that although broadcast TV news coverage of environmental justice was more common in 2021 than in the previous four years, it remains scant. That year, broadcast news shows on ABC, CBS, and NBC aired a combined 153 segments about environmental impacts, regulations, or health hazards, but only 19 of those segments (12 percent) included mention of Black and Brown communities most affected by climate and environmental degradation.[13] In the year 2021, mainstream broadcast networks' meager attention to issues of justice suggests that although our media environments are radically changing, media content and frames, especially in legacy outlets, in some fundamental ways are slow to evolve.

Networked Publics

The late 1990s and early 2000s marked the beginning of a transformative moment as internet access was becoming ubiquitous and people were experimenting with their newfound ability to produce rather than just consume media. Indeed, a less centralized and more democratic media environment seemed possible.

In 2002, I was part of a group of scholars at the University of Southern California focused on exploring the impacts that transition from mass media to networked media might have on publics. A passage from the resulting publication, *Networked Publics* (2008), sums up the largely optimistic view we brought to our study:

> The convergence between old and new media is tied to broad-based changes in how power and information are distributed across society, geography, and technology.... Through the Internet, casual communication, personal stories and opinion, homegrown news and amateur cultural works can be made easily available to large audiences. As a consequence, the top-down, one-to-many relationship between mass media and consumers is being replaced, or at least supplemented, by many-to-many and peer-to-peer relationships.... With individuals rising in influence through news blogs and other emerging forms of communication while old media lose influence, questions of who has authority come to the fore.[14]

Our confidence in the power of networked technology to upend the top-down power dynamics of the mass-media era built on the work of scholars such as Yochai Benkler, who theorized that we were at the beginning of a shift away from commercial media and centrally organized knowledge production and toward "non-market" and distributed production.[15]

Networked technologies and infrastructures had the potential to democratize journalism in terms of what stories were told, by whom, and from whose perspective. Deprofessionalization meant more opportunities to get the public involved. Whereas in some industries such as marketing and gaming, opening the floodgates and letting the public in was proving problematic, journalism studies and networked-public scholars imagined journalists eagerly adopting these new tools and platforms and leveraging them to do their work in ways that would reaffirm their journalistic authority.[16] This optimism carried well into the second decades of the 2000s. According to Seth Lewis and Logan Molyneux in 2018, "The collective hope for social media and journalism over the past decade, as painted especially in the trade press but also in the academic literature, has been one of implicit positivity: that, on balance, social media would be a net benefit for individual journalists, for journalism as an institution, and for society as a whole."[17]

For example, at that time journalists turned to blogs and later to sites such as Facebook and Twitter to find sources for their stories, to develop more direct connection with their readers, and even to enlist them as helpers.[18] Low-cost tools and accessible networks of distribution fueled citizen journalism, and advocacy and activist groups' adoption of practices traditionally reserved for reporters expanded the field of journalism beyond traditional newsrooms. Networked publics used affect as political statements and in the process engaged both politically and socially, "feeling their way into" news issues and events, shaping narratives, and at times invigorating dynamic public engagement.[19] Early in this transformative moment, Zizi Papacharissi zeroed in on the emotion that characterized much social media communication and documented the use of affect by networked publics in various contexts, including the Arab Spring protests, the Occupy Wall Street movement, and everyday personal expressions, in order address the question "What is the form that publics take as they are called into being through the connective structure of feeling?"[20]

The answer, as we have seen, is that it very much depends. Networked technologies have certainly opened access to more diverse forms of participation and types of participants. Affective publics have been tremendously engaged and impactful—for instance, in response to the killing of Black Americans by the police (#BlackLivesMatter), to widespread sexual abuse and assault (#MeToo), and to the escalating climate crisis. Although these earlier positive assessments of the potential of digital tools and platforms to enhance journalism and public engagement were not wrong, they failed to anticipate a large part of the story.

Networked technologies have unleashed a flood of feelings into political discourse, and the problem is that whether they are emotional appeals to publics or responses to these appeals or campaign messaging or bot-generated missives, they often are treated as more compelling evidence than facts (James Inhofe's snowball is a case in point).[21] Today, alarm bells everywhere are sounding warnings about our networked infrastructures' impact on public engagement in democracy,[22] highlighting what has been alternatively called "the automated public sphere,"[23] "the unedited public sphere,"[24] a state of utter "information disorder,"[25] "disrupted public spheres,"[26] a truthless public sphere by design,[27] or a toxic waste dump.[28] More specifically, the networked-publics research group in which I participated and others like it failed to anticipate the impact of an intensifying epistemic rift—the loss of common notions of truth—in combination with the rise of big tech.

Automated Publics

Platforms and search engines, which include Google, Facebook, YouTube, Twitter, Instagram, Snapchat, TikTok, among many others, host, store, and serve up content for public circulation.[29] They do not produce or commission the content but rather make important choices about where and to whom content generated by users will be distributed and who won't ever see it.[30] They do all of this through the mass collection of data about its users.

The idea that data are a by-product of political, economic, and social activities is by no means new. The amount and value of data collected today, however, have vastly increased, fueling development and advancement in so-called big data—large data sets that can be analyzed to find patterns in users' interests and behaviors in order to predict future interests and behaviors and what content will be of most interest to users. There is indeed growing evidence that the collection and use of big data do not merely measure reality but also shape it[31] and that we have in turn become digital citizens. What makes us digital citizens is not our actions but the digital traces we leave behind when we do just about anything—our phones collect geolocation data, our online communications leave metadata, our likes and clicks provide data on our preferences, our activities in smart cities and homes are traced by sensors.[32] Shoshana Zuboff, professor emerita at Harvard Business School, describes this economic system centered around personal data as "surveillance capitalism,"[33] which is built on a logic that promotes data as a stand-in for the public. Assuming that our digital footprints add up to the sum of who we are, however, is both erroneous and harmful.

Prediction and Control

Under conditions of surveillance capitalism, those who collect, manage, analyze, and control personal data have unprecedented insights that they can use to predict and control people's activities, which Arne Hintz, Lina Dencik, and Karin Wahl-Jorgensen argue presents us with a new and insidious power dynamic "premised on an order of 'haves' and 'have nots' between those who provide personal data (digital citizens) and those who own, trade, and control it (typically, large Internet companies and the state)."[34] Tech companies track media consumption and cross-reference this information with platform users' personal interests, demographic information, and purchasing history. They then aggregate, consolidate, and sell user data to content producers—or anyone—seeking a robust view of their audience's habits, tastes, and weaknesses. Governments and corporations use these data to compartmentalize us (users) according to our political preferences and activities, our consumer habits, and our likelihood of committing a crime or being involved in a protest, which, in turn, affects the type and quality of services that we may receive or the ease with which we may move freely within and across national borders.

Climate activists, lawyers, and journalists, for example, are often the subject of online government surveillance. In a Swedish Society for Nature Conservation survey in 2018, nearly 70 percent of environmental organizations in the Global South reported having been subject to physical and digital surveillance meant to gather information for the purpose of predicting and countering their efforts or intimidating and punishing climate advocates.[35] In the United States, law enforcement authorities partner with private-security companies to surveil activists in an attempt to anticipate and control protest activities.[36] Evidence of this collaboration was widely documented in 2017 during the protests against the Dakota Access Pipeline at the Standing Rock Indian Reservation.[37] A Freedom of Information suit brought by

The Guardian in 2018 revealed that the FBI identified 350.org—the climate-activism group founded by Bill McKibben—as a terrorist campaign and both physically and digitally tracked its members.[38] Classifying peaceful protestors and organizations as terrorists is a tactic commonly used by the U.S. government to discredit and strip them of their legal protections.[39]

In addition to government tracking activists to predict their behavior and counter their efforts, predictive policing, for example, uses data on neighborhood crime rates, residents' previous crimes, personal living conditions, and a range of other characteristics to create categories of potential future criminals.[40] Credit scores and insurance rates increasingly depend on monitoring and processing personalized data. The most comprehensive plan built with the social media credit-score system rolled out in China in 2020, which draws on social media activities, friends, messages, spending habits, and other data to determine what level of services different categories of citizens should enjoy access to.[41] Safiya Noble calls this practice "technological redlining," referring to the face-to-face discriminatory processes historically tied to race whereby financial and other services are denied to potential customers based on the area where they live. "We have new models where financial institutions are looking at our social networks to make decisions about us.... [I]f we have too many people who seem to be a credit risk in our social networks, that actually impacts the decisions that get made about us." Noble further points out that whereas we used to be able to identify and possibly hold accountable a racist banker, these new ways of categorizing people preclude the possibility of calling anyone out because they fall outside existing categories of civil rights, are created without our knowledge, and remain largely invisible, which makes them especially insidious.[42]

Isolation

Data are also used to isolate users by tailoring content, or "micro-targeting," according to what is presumed to be our preexisting values and beliefs, and AI is engaged to automate what shows up in our feeds. In media production, the use of big data signals a massive shift toward a much greater ability to find audiences for advertisers and a greater flexibility in tailoring advertising and media content toward personal tastes, depending on whether executives in particular media sectors might (or might not) consider audiences present or future valuable users of their materials.[43] To ensure that we see content we agree with or will likely find interesting, algorithms dictate what news and other online content becomes visible or remains invisible to us—or, in Arendt's phrasing, what appears to us. By doing so, they manage and manipulate the public realm—how the world, other people, their stories, and action appear to us.

It is well documented that the fossil-fuel industry uses social media to spread misinformation. For example, industry-supported climate-change-denying groups are using Facebook's advertising platform and unique targeting abilities to intentionally seed doubt and confusion around the science of climate change. Research by InfluenceMap, a London-based nonprofit that tracks climate lobbying, identified fifty-one climate disinformation ads, which were viewed an estimated 8 million times during the first half of 2020. These ads were targeting mostly older men in rural areas. Facebook took down just one of the fifty-one ads identified by InfluenceMap. These ads were run by well-known groups that collectively have a total revenue of around $68 million per year.[44] Indeed, Facebook was the go-to place for denial groups to spread disinformation to millions of people in the lead-up to the 2020 U.S. elections in part because it so effectively targeted these ads to isolated groups of people who would be receptive to their message.

The downsides of this tech-induced isolation received a great deal of attention when the Cambridge Analytica scandal in 2016 revealed that the company harvested the personal Facebook data of 87 million users in order to manipulate how they vote.[45] Personal data were used to "microtarget" different groups with election-related messages, including massive efforts to dissuade people from voting, as well as with racist and anti-Semitic messaging. With the ability to secrete their messaging away, political operatives were emboldened. The story made plain the ways data-collection and sorting practices have transformed election campaigns and other political communication strategies through the use of psychometric data to microtarget advertisements, amounting to covert and deceitful messaging.[46]

One of the most commonly held ideas about the internet and about microtargeting in particular is based on early research that points to the existence of echo chambers. As the story goes, large swaths of the public are overexposed to news and information that fit with their preexisting views,[47] thus forming tight networks of like-minded people or echo chambers that serve as reassurance that one's opinions are sound and that divergent opinions are not.[48] More recent research, however, has demonstrated that people *are* exposed to differing opinions online—that the filter is not as effective as scholars initially feared—but that this exposure makes them even more committed to their original perspective.[49] Research also suggests that the groups most vulnerable to disinformation online are like-minded users who share that disinformation among themselves rather than vicious outsiders trying to fool unwary victims. In particular, political microtargeting on social media can and is being used to manipulate vulnerable groups *from the inside*, which serves to trigger and continually reinforce social division.[50] So the problem with the narrowness of what appears to us online is not just that it is narrow but also that it is easily manipulatable. Indeed, isolation, even if not absolute, makes us vulnerable.

Attention

Those who collect, manage, analyze, and control personal data not only predict and isolate but also have a monopoly over the digital advertising market and dominance over the attention that these data secure. As Herbert Simon famously wrote in the 1960s, "In an information rich world, a wealth of information means a dearth of something else: a scarcity of whatever it is information consumes. What information consumes is rather obvious: it consumes the attention of its recipients. Hence a wealth of information creates a poverty of attention and a need to allocate attention efficiently among the overabundance of information sources that might consume it."[51]

In the current context, we can think of journalism as being adversely affected by an overlapping poverty of revenue and attention. Big tech's dominance of the advertising market undermines the financial viability of journalism, a fact that was explicitly addressed in a U.S. Senate hearing in July 2020. In her opening statement, Senator Maria Cantwell of Washington State said, "Local news media have lost 70 percent of their revenue over the last decade, and we have lost thousands, thousands of journalistic jobs that are important. . . . Somehow, we have to come together to show that the diversity of voices that local news represents, needs to be dealt with fairly when it comes to the advertising market. And that too much control in the advertising market puts a foot on their ability to continue to move forward and grow in the digital age."[52]

Cantwell's concern was raised by a series of newsroom layoffs and highly publicized discussions of the role of big tech in journalism's collapse. When Vice Media Group announced the layoff of 155 employees worldwide in May 2021, CEO Nancy Dubuc pointed a finger at big tech: "We grew our digital business faster than anyone at a time when we believed that as more pies were baked, we'd keep getting a slice. But we aren't seeing the return from the platforms benefiting and making money from our hard work. Now, after many years of this, the squeeze is becoming a chokehold. Platforms are not just taking a larger slice of the pie, but almost the whole pie."[53] Added to this financial squeeze, journalism now must compete for attention within a vast sea of content, much of which is created with the sole aim of garnering attention rather than educating the public or facilitating dialogue. Thus, the more extreme the content, the more likely it will garner attention. The more people engage with that content, the more data extracted from users and the more advertising dollars made by big tech.[54] Journalists are thus faced with mounting pressure to create content that can compete online.

THE MYTH OF BIG DATA

Nick Couldry points out that these transformations go beyond the impact of big tech on journalism and even beyond the sorts of content that are privileged online. In his inaugural London School of Economics lecture in 2013, he discussed "myths" that we are adopting—both inside and outside the academy—to make sense of the transformation taking place in the public sphere(s), warning that the "myth of us" paints the picture of social media users as a "natural collectivity"; "We must be wary when our most important moments of 'coming together' seem to be captured in what people happen to do on platforms whose economic value is based on generating just such an idea of natural collectivity. 'We,' the collectivity of everyday people, everywhere. Vague as it is, this claim grounds any number of specific rhetorics and judgements about what's happening, what's trending, and so (by a self-accumulating logic) what matters: for government, society, business, and for us."[55] The value of this new social domain, Couldry pointed out, is inextricably linked to the "myth of big data," or the idea that computer-based analysis of large data sets can offer a new route to understanding the social, both for scholars doing research on the public sphere and for platforms aiming to financially exploit these insights.[56]

The science and technology studies scholar Noortje Marres describes the myth of big data like this: "Computational social scientists prize online platforms for enabling a science of society that does without interpretation: on Facebook and Twitter, the argument goes, social and political processes can be measured by tracing action—by what is shared, linked, clicked, purchased[,] . . . [and they] consider[] it progress that in digital media public opinion formation does not need to be defined as an interpretative process."[57] In a striking example of how this myth has been inserted into public discourse, the *Wired* editor Chris Anderson wrote in the article "The End of Theory: The Data Deluge Makes the Scientific Method Obsolete" (2008), "This is a world where massive amounts of data and applied mathematics replace every other tool that might be brought to bear. Out with every theory of human behavior, from linguistics to sociology. Forget taxonomy, ontology, and psychology. Who knows why people do what they do? The point is they do it, and we can track and measure it with unprecedented fidelity. With enough data, the numbers speak for themselves."[58] The version of society that big data generate, according to Anderson, is "good enough." But good enough for what?

Tech companies' use of big data is certainly good enough for them to turn massive profits. But as Melissa Aronczyk points out, the data collected are often relevant only to specific behaviors carried out in a predetermined context. They reflect behaviors shaped less by user interests, concerns, or habits than by the way the system is engineered to maximize attention and profit and to surveille while also adhering to security and privacy regulations. These factors, Aronczyk argues, create data sets that are highly specific to the conditions of their collection and that cannot, indeed should not, stand in for any public.[59] Chris Bail, a computational sociologist who studies so-called echo chambers, concurs, arguing that social media do not reflect society like a mirror but rather refract it like a prism, splintering our identities into artificial categories, "leaving us with a distorted understanding of each other, and ourselves." He writes, "The social media prism fuels status-seeking extremists, mutes moderates who think there is little to be gained by discussing politics on social media, and leaves most of us with profound misgivings about those on the other side, and even the scope of polarization itself."[60] That is, social media data are highly suspect as a stand-in for the social world.

NOISE, INCIVILITY, AMBIVALENCE: FACTORS THAT SHAPE HOW WE COMMUNICATE ABOUT THE CLIMATE CRISIS

All these issues point to the fact that journalists, no matter how stellar their work, work on a vastly uneven playing field and are often drowned out by extreme content or given access only to content generated by those who are already interested and informed about climate issues. Three characteristics of online communication about climate—noise, incivility, and ambivalence—pose challenges to those who seek to engage in meaningful online discussion and action. Not exclusive to communication on climate, these three characteristics illustrate more broadly the challenges to online information and discussion around topics in which powerful entities have a vested interest, people strongly identify with their political position, and companies want to exploit the passions and outpouring of attention associated with online activity.

Noise

Journalistic climate coverage, no matter how compelling or high quality, has to compete with a sea of other content online, and it often gets drowned out in that competition. Recent studies of online communication on climate change have demonstrated that traditional journalism is less popular than user-generated content online, despite the fact that user-generated content is often inaccurate.[61] The science communication scholars Dustin Welbourne and Will Grant found that although professional media outlets post far more content, amateur science videos on YouTube gain vastly more attention.[62] The majority of climate-related amateur content online opposes scientific consensus or claims that there is still significant scientific debate on the fact of anthropogenic climate change.[63] A study by Joachim Allgaier in 2019 analyzed the content of a randomized sample of 200 YouTube videos related to climate change. He found that 107 of those videos either denied that climate change was caused by humans or claimed that climate change was a conspiracy.[64]

The digital culture scholars Bernhard Rieder, Ariadna Matamoros-Fernández, and Òscar Coromina point out that YouTube arranges search ranking in a way that allows frequently posting users with fringe perspectives to gain exceptional levels of visibility. They write, "Feeding on controversy and loyal audiences, these channels consistently appear in top positions, even if their videos most often receive fewer views than more mainstream or conciliatory voices."[65] This practice of privileging extreme points of view extends to the whole online landscape and is central to tech companies' business model. Zeynep Tufekci and others have documented the way recommendation algorithms and other forms of everyday AI technologies create an architecture of persuasion: view a climate video that gives space to climate skeptics and watch the recommendations pile up for similar videos in increasing order of extremism—videos set to autoplay for your convenience, a YouTube spigot of climate-change-denial propaganda.[66]

So the challenge is not only that amateur and false information is more popular but also that an environment that prioritizes attention-getting "virality" fuels efforts to manipulate public discourse, "regardless of whether it is true or even minimally decent," as Frank Pasquale puts it.[67] More often than not, the content that garners the most attention is some combination of the emotional, the extreme, and the controversial. Frequency of user reactions in turn influence how visible the content is in other people's feeds.

As elaborated in chapter 1, a well-funded and persistent flow of disinformation about climate science originates from right-wing corporations, think tanks, legislative groups, and activists.[68]

These efforts leverage various sorts of media, but they are particularly pernicious when deployed on social media platforms to rapidly spread false and deceptive information. As fires raged in Australia in early 2020, rumors spread that they were the work of left-wing arsonists intent on creating panic and support for their cause rather than the result of a climate change or other environmental factors. The conspiracy theory originated on Twitter and was promulgated by bots—automation tools programmed to post autonomously and to mimic human behavior online—using the hashtag #ArsonEmergency.[69] When massive wildfires broke out in the western United States later that year, the conspiracy theory continued to spread, this time with humans rather than bots playing a central role. The Facebook group Law Enforcement Today posted an unsubstantiated rumor that members of "antifa"—an antifascist group associated with the far political Left in the United States—had started the wildfires. The post was shared 7,100 times before Facebook finally took it down, but by then the harm was already done. Efforts by local fire and police agencies to respond to the fires were disrupted as armed vigilante patrols roamed towns gripped by raging fire rumors; journalists and others were attacked; phone lines were clogged with baseless claims of antifa looting and arson; and people refused to evacuate in an attempt to stand their ground against the alleged arsonists.[70] In a statement released on September 11, 2020, FBI Portland and local law enforcement agencies implored people to stop the spread of disinformation: "Conspiracy theories and misinformation take valuable resources away from local fire and police agencies working around the clock to bring these fires under control. Please help our entire community by only sharing validated information from official sources."[71]

The spread of disinformation, whether by bot or human, is meant to make noise and disrupt other conversations. In the case of the wildfires, it drowned out discussions of the link between climate change and the fires. In 2019, following a CNN Climate Crisis Townhall in September featuring Democratic Party presidential primary candidates, noise was used to drown out their discussion. Just after the event, *climate change* became a top-trending topic not because of an organic surge in interest in discussing the topic but as a way to preemptively disrupt that potential discussion. Bot Sentinel, a tool that tracks automated accounts, reported an unusually high number of mentions of the term *climate change* (seven hundred) in a twenty-four-hour period from some one hundred thousand Twitter accounts. Christopher Bouzy, founder of the Bot Sentinel platform, explains, "When a topic like climate change trends among the trollbots, it is likely there is some amount of coordination involved. What we are noticing is that these phrases are more than likely being pushed by accounts that have an agenda."[72] Indeed, the surge of activity set off by the townhall was aimed at ridiculing the Democrats and dismissing any serious discussion of climate science on Twitter. Accounts began sharing video montages of the event edited by Trump's reelection campaign and by the far-right web outlet the Gateway Pundit to portray the candidates in the worst possible light.

Although in this instance the action was likely driven by U.S. political operatives, the introduction of noise into public discourse is also a common tactic of foreign governments intended to stoke domestic tensions. For example, in 2017 an Instagram account called "@Native_Americans_United" attempted to exacerbate political friction around climate change by sharing images related to Native American social and political issues, including the construction of the Dakota Access Pipeline across the northern Plains and midwestern states. The account was one of 180 accounts connected to a Russian troll farm deployed to exploit tension points in American politics, such as elections, police violence, and climate change.[73]

These tactics and the infrastructures that support them are sophisticated, ever-evolving, and powerfully disruptive to widespread understandings of climate-change realities and impacts.

Incivility

Beyond intentional disruption through bot- and troll-produced noise, journalists and their sources who acknowledge the existence of climate change are subject to an exceedingly inhospitable environment where they are regularly threatened and harassed. Incivility is especially prevalent for women and members of Black and Brown communities, who face sexist and racist attacks on a routine basis online.[74]

Violence against and harassment of scientists and journalists in general and of those focused on the climate crisis in particular are also on the rise.[75] Covering the environment has become one of the most hazardous beats in journalism because the reporting of environmental threats often involves business and economic interests, political battles, crime and corruption, and high-stakes struggles over Indigenous rights to land and natural resources.[76] It is also hazardous because the environment has become a site of heated identity politics. With regard to contested issues such as climate change and COVID-19, for example, scientific proof is very rarely at stake.[77] As Candis Callison puts it, "Instead, science becomes a factor and/or catalyst in often boisterous debates revolving around political, moral, and ethical claims."[78]

These contests over what constitutes truth and an increasing mistrust of both scientists and journalists contribute to online and offline hostility. Even in countries that have traditionally been considered safe and supportive of press freedoms, journalists have to contend with political climates where media are distrusted and thus become prime targets of hostility.[79] As Seth Lewis, Rodrigo Zamith, and Mark Coddington point out, online harassment can lead even to more aggressive abuse offline.[80] This is true for journalists as well as for high-profile scientists, politicians, and activists who speak out on climate issues.[81]

Katherine Hayhoe is one of many climate scientists, activists, and politicians who routinely faces personal threats. She says she receives hundreds of emails and letters after her media appearances, calling her a fraud and a liar and threatening her life and the lives of her family. "There are people who become dedicated to following you, who have Google alerts set up on your name, who stalk your Twitter and Facebook accounts, who essentially make a career out of ridiculing and vilifying you," she said in 2015.[82] The climate activist and youth activist leader Greta Thunberg faces a near-constant barrage of online attacks. Soon after she launched a two-week journey by sailboat across the Atlantic in the fall of 2019, for example, Arron Banks, the British political donor and cofounder of the Brexit Leave EU campaign, tweeted, "Freak yachting accidents do happen in August."[83] Minister of Environment Catherine McKenna of Canada was assigned a security detail in 2019 after a car pulled up alongside her and her children, the driver cursing her and calling her a "climate Barbie"—a name that Gerry Ritz, a Canadian Conservative member of Parliament, first called her on Twitter during an online exchange over a climate-change report.[84]

The rancor aimed at Hayhoe, Thunberg, McKenna, and so many others for speaking out about climate change reflects not only a deeply fractured political environment but also infrastructures that encourage such rancor. One study done in 2013 demonstrated what its authors called the "nasty effect" of uncivil online comments. The 1,183 participants were asked to read and comment on a blog post that neutrally presented the risks and benefits of nanosilver, tiny particles of silver used mostly for medicinal or antibacterial purposes. Half of the subjects were then exposed to civil comments and the other half to rude ones that said things such as "If you don't see the benefits of using nanotechnology in these kinds of products, you're an idiot." They found that uncivil comments both polarized readers and changed their interpretation of the post itself, concluding: "Perceptions towards science are shaped in the online blog setting not only by 'top-down information,' but by others' civil or uncivil viewpoints, as well. While the Internet opens new doors for public deliberation of emerging technologies, it also gives a new voice to non-expert, and sometimes rude, individuals."[85] With an estimated 60 percent of Americans saying that the internet is their primary source when looking for information about scientific matters, nasty reactions can have enormous influence in shaping what we know or what we believe about science.[86]

This uncivil environment compromises the ability of publics to exchange information and engage with climate issues both online and offline. It constitutes another barrier to open and robust discourse about the climate.

Ambivalence

A general ambivalence permeates the internet on all levels, which is particularly evident in the case of communication about the climate crisis. By saying that the internet is "ambivalent," I mean that it is capable of being used to both help and harm, to elicit both joy and anger, and both to facilitate robust and diverse discussion and to hamper that same type of discussion. I am building here on the work of Whitney Phillips and Ryan Milner, who use the concept of ambivalence to emphasize the positive and negative potential of online cultural expression rather than to refer to indecision ("I could go either way"; "I'm ambivalent about which movie to watch") or ambiguity ("I'm not sure what she means"; "her message was pretty ambivalent").[87]

Many contemporary theories that shaped initial assessments of the internet—"peer production,"[88] "convergence,"[89] "produsage,"[90] and, perhaps especially, "networked publics"[91]—were based on the assumption that these practices were driven by earnest people working to create public goods such as information, knowledge, and positive political action. Indeed, failure to anticipate the detrimental impact of this new era of networked media infrastructure was at least in part due to what once was the belief in an earnest internet or, broadly speaking, communication scholars' assumption that people act in good faith, care about facts, behave in a way that resonates with their values.[92]

Phillips and Milner argue instead that in the context of digital culture we cannot assume people's intentions or motivation and suggest that how people behave online through "folkloric expression of vernacular creativity," namely hashtags and memes, can be used to promote social good or social ill. They write: "[Such online expression] can 'hijack identities' causing immense personal distress as well as cement collective and collaborative relationships that are both pro- and anti-social, often at the same time."[93]

Early assessments of the internet and social media erroneously assumed earnestness not only on the part of publics but also on the part of the tech companies, yet these companies have reshaped communication infrastructures based not on public service, such as the building of public roadways, but rather on the relentless drive for profit. That is, we had high hopes that networked technologies would transform us into newly activated and visible digital citizens and that platform infrastructures would support this shift. But we are instead left shouting to be heard over the noise and conducting public discourse on platforms that have no incentive to improve the quality of their service to the public because service to the public was never where their commitments lay.

A performative earnestness was on display in 2018 when Mark Zuckerberg was first called to answer questions from the U.S. Congress about Facebook's failures during the 2016 election—including the selling of ads to Russian propagandists and the permitting of disinformation to spread fast and wide on the platform. In a prepared statement, Zuckerberg wrote: "It was my mistake, and I'm sorry. I started Facebook, I run it, and I'm responsible for what happens here. So now we have to go through every part of our relationship with people and make sure we're taking a broad enough view of our responsibility."[94]

Much of the questioning from the Senate and House revolved around asking Zuckerberg if he would commit to doing this or that or what he thought should be done. There was a sense that because he seems like a good guy—polite and apologetic—we should leave it up to him to fix Facebook and do the right thing.

But consider in this light the infamous Zuckerberg quote from 2003 during the first weeks of Facebook (then called Facemash), when he was working to assure fellow students at Harvard that they could trust him with their personal info. A private instant-messenger conversation from that time was leaked and published by Silicon Alley Insider.

> ZUCK: yea so if you ever need info about anyone at Harvard
> ZUCK: just ask
> ZUCK: i have over 4000 emails, pictures, addresses, sns
> FRIEND: what!? how'd you manage that one?
> ZUCK: people just submitted it
> ZUCK: i don't know why
> ZUCK: they "trust me"
> ZUCK: dumb fucks[95]

The point is not that Zuckerberg in particular is not to be trusted but rather that we cannot and should not rely on the goodwill of tech companies or their founders to create infrastructures that support the public sphere. If we consider ambivalence rather than earnestness as the orienting ethos of platforms, we can better take account of the forces that contribute to the erosion of social trust. As Ashley Hedrick, Dave Karpf, and Daniel Kreiss suggest, "This [erosion] goes far beyond the loss of trust in journalism or even institutions; it cuts to the heart of everyday social relations and public discourse."[96] We can see how this ambivalence—or the ability both to create harms and to provide solutions—plays out by looking at the contradictory ways Facebook addresses climate issues.

In September 2020, Facebook launched the Climate Science Information Center, a Facebook page for climate-change information, curated by what Facebook identifies as "a team of experienced journalists at Facebook" and modeled on the platform's COVID-19 Information Center, launched earlier that year. The page reads, "All data comes [sic] from the

Intergovernmental Panel on Climate Change (IPCC), the United Nations body for assessing the science related to climate change. Wording is edited for length and clarity."[97] With this initial launch, the page was available in the United States, the United Kingdom, Germany, and France, with plans to expand to other countries in the coming year. Facebook also announced plans to achieve net-zero carbon emissions for its global operations by 2020 and to reach net-zero emissions for its entire value chain by 2030.[98] The "solutions" section of the page recommends individual-focused solutions such as turning off the lights, washing your clothes less, and recycling. There is no mention of the more egregious causes of climate change or the fact that just one hundred companies are responsible for 71 percent of emissions, among other salient information.[99] There is no mention of the fact that the tech industry's carbon emissions account for 2 to 3 percent of global emissions, according to the UN.[100]

The climate journalist Brian Kahn sees these efforts by Facebook as "obscur[ing] the systematic changes needed to address climate change while peddling tips that mirror Big Oil talking points. The whole thing is a giant hand-wave to distract us from looking at the real solutions to climate change and the role Facebook is playing in corroding them." He goes on, "What Facebook is doing is akin to the National Rifle Association's argument that the only way to stop a bad guy with a gun is a good guy with a gun. In this case, Facebook is the arms dealer handing out guns to both sides."[101]

Indeed, Facebook's weak climate-related efforts must be measured against the fact that climate-change denial and conspiracy theories run rampant on the platform, sometimes in the form of paid advertising, as InfluenceMap documented. This proliferation is not an accident but the result of company policy and practice that supports the flow of disinformation.

In 2018, in an attempt to curtail political manipulation and increase transparency, Facebook announced it would require organizations placing political and issue ads on the platform to register them as ads and include a "paid for by" label in each ad. The announcement drew a lot of coverage, but the announced measures and other content-moderation efforts failed to curtail the spread of disinformation. The next year, under pressure to improve content moderation, Facebook partnered with several fact-checking outfits, one of which turned out to be CheckYourFact.com, an arm of the right-wing news outlet *The Daily Caller*, funded in part by the oil barons Charles and David Koch, who have long funneled money to groups who spread misinformation on climate change.

To many, the influence of climate deniers on Facebook's content-moderation practices is clearly evident—climate scientists, activists, and journalists regularly see their posts flagged as political ads and blocked. In 2019, the climate scientist Katharine Hayhoe tried to share a post on Facebook about hurricanes and climate change that read, "'Was it caused by climate change?' is the most common question we hear about an extreme event. But when it comes to hurricanes, that's the wrong question. The right one is, 'How much worse did climate change make it?'" Facebook flagged her post as an unauthorized political ad and blocked it.[102]

But while posts by climate scientists are labeled ads and blocked, climate disinformation is labeled "opinion" and escapes the scrutiny of fact checkers. In 2019, climate scientists working as fact-checkers for Facebook flagged and marked as "false" a post by CO_2 Coalition, a group dedicated to promoting the idea that the world needs to burn more fossil fuels and funded by groups that oppose regulations on fossil fuels. A Facebook employee removed the label "false" and changed it to "opinion," which allowed it to be shared.[103]

Caleb Rossiter, the executive director of CO_2 Coalition, says he regularly uses Facebook to reach an audience outside of conservative media: "It's a huge reach. You can reach so many people both with your posts and your advertisements. . . . We're kind of like Donald Trump. We're not happy with the treatment we're getting from the mainstream media, we resort to social media. That's where our action is in larger part." The coalition reaches people via Facebook content labeled as opinion pieces as well as through ads, one of which sums up its mission: "We are saving the people of the planet from the people who claim they are saving the planet." Those ads have received more than fifty thousand impressions.[104]

What accounts for these contradictory uses of Facebook—on the one hand to provide quality science-based information through the Climate Science Information Center but on the other to provide a platform for the rampant spread of false information? Shoshana Zuboff explains the lackadaisical efforts at content moderation:

> Asking a surveillance extractor to reject content is like asking a coal-mining operation to discard containers of coal because it's too dirty. This is why content moderation is a last resort, a public-relations operation in the spirit of ExxonMobil's social responsibility messaging. In Facebook's case, data triage is undertaken either to minimize the risk of user withdrawal or to avoid political sanctions. Both aim to increase rather than diminish data flows. The extraction imperative combined with radical indifference . . . produce systems that ceaselessly escalate the scale of engagement but don't care what engages you.[105]

She references as evidence a leaked memo in which Facebook executive Andrew Bosworth describes the company's intentional disregard for truth: "We connect people. That can be good if they make it positive. Maybe someone finds love. . . . That can be bad if they make it negative. . . . Maybe someone dies in a terrorist attack. . . . The ugly truth is . . . anything that allows us to connect more people more often is *de facto* good."[106] This is the essence of ambivalence.

This ambivalence has spurred activists to turn their attention to platforms. Climate Creatives (a nonprofit association of creative professions), Greenpeace, and the youth civic-engagement group Hip Hop Caucus are leading a coalition of climate organizations calling on the CEOs of Facebook, Google, Twitter, LinkedIn, and TikTok to ban fossil-fuel advertising from their platforms. Anusha Narayanan, Greenpeace's U.S. climate-campaign manager, explains: "The definition of hypocrisy is social media giants saying they care about environmental impacts while accepting millions of dollars from fossil fuel corporations to peddle their propaganda."[107]

In the United Kingdom, the Conscious Advertising Network is trying to ensure ad industry ethics is updated to reflect the contemporary realities of the technology behind advertising today. Harriet Kingaby, the network's cochair, describes how this update pertains especially to climate information. She says that the internet has brought about mass data collection, targeted ads, and enormous amounts of ad space (a.k.a. "ad inventory"), factors that are "turbocharging climate disinformation." Advertisers are the major driving force behind platforms. And because platforms are the new gatekeeper, promoting the most salacious content and syphoning us off into like-minded groups, advertisers are complicit in creating this toxic environment by inadvertently paying to support it.[108]

WE WANT OUR FACTS BACK!

In response to an information environment characterized by noise, incivility, and ambivalence, there is widespread nostalgia for a time when the elite authority of experts went unquestioned and fact-based evidence was seen as an essential part of public debate.[109] This nostalgia, though, overlooks the fact that when experts ruled, many people were excluded from the public conversation. It also ignores the fact that policies based on a narrowly defined expertise and elite authority bypass different lived experiences and often fail to lead to sustainable solutions. Knowledge about the world is also crafted by morality, beliefs, personal experience, emotional work, and social action. Essentially this pining for the past is based on an amnesia for the days in the more recent past, the 1990s and early 2000s, when there was hope that the internet would provide a space for diverse voices and robust dialogue. It's based on forgetting that not long ago we wanted to challenge the exclusive dominance of elite experts.

Based on a desire to return to the days when elite authority reigned, fact-checking and media-literacy initiatives—often led by journalists—seek to correct and slow the flow of disinformation, differentiate responsible and irresponsible users, and know or unknow subjects; these initiatives mirror broader conflicts between educated progressives and by and large less educated supporters of populist leaders and movements at the center of contemporary debates about disinformation. That is, they position potential populists as the drivers of epistemic crisis that need to be contained, whether through content moderation, fact-checking, or literacy campaigns that insulate the "good publics" from the populists' corrosive potential. What they do not do is account for the role platforms play in undermining respect for knowledge in the first place.

The absence of regard for facts online goes beyond the quality of users and sources of information and instead is rooted in digital infrastructures and how they shape the conditions for public discourse. It includes the way data are collected, content is privileged, and connections are enabled and disabled. This lack of regard for facts also has to do with a fundamental premise that situates knowledge about the social world as separate from the actual people it represents. Returning to the "myth of big data," or the idea that computer-based analysis of large data sets can offer a new route to understanding the social without actually considering social actors, underscores the point that social media are not indifferent to knowledge but that the knowledge they produce reflects platforms' economic demands instead of truth-based realities.[110] Noortje Marres points out that "this analytic architecture is shot through by behavioral assumptions: the activities that platforms enable—to influence, to make trend—have the 'manipulability' of users as their primary feature."[111] This feature makes the fidelity of the "knowledge" produced problematic because it doesn't matter if the producers of information are right or wrong, only that they render the subject manipulatable. As discussed earlier, these proxies generated by big data depict its data subjects in ways that don't connect anymore with the space of action and thought in which actual individuals live. Thus, misrepresentation and disinformation begin long before they manifest downstream in content that can be fact-checked.

What can we find out by looking at actual spaces of action and thought? How are activists and publics engaging with the climate crisis in ways that create meaning through lived values and experience? The next two chapters address these questions and others related to how people are responding to the simultaneous climate crisis and information crisis, and they offer a blueprint for media narratives, material conditions, and expanded conceptions of authority that will foster rather than preclude a way forward.

3

AFTER PEAK INDIFFERENCE

When society faces a horrific but slowly intensifying problem, we tend to be indifferent to it until it gets right in our face. "Peak indifference" is what Cory Doctorow calls the moment a seemingly far-off problem becomes so obvious people feel alarmed and start paying attention.[1] With the rise in frequency and intensity of extreme heat and cold, hurricanes, drought, and wildfires, more and more people are experiencing firsthand the adverse effects of the climate crisis.

The Yale Program on Climate Change Communication and the George Mason University Center for Climate Change Communication survey on American attitudes about climate change found that in 2020, a year in which there were twenty-two separate billion-dollar weather and climate disasters across the United States,[2] 60 percent of those surveyed said they thought that climate change was caused by human activity, up from 42 percent in 2013, and 73 percent thought global warming was happening, up from 63 percent in 2013.[3] In 2021, a Pew Research Center survey in seventeen countries with advanced economies in North America, Europe, and the Asia-Pacific region found widespread concern (72 percent somewhat or very concerned) about the personal impact of global climate change and widespread willingness

(80 percent) to make changes to how they live and work to reduce the impacts of climate change.[4] We are, it seems, no longer collectively indifferent to the climate crisis.

Why indifference has persisted for so long, especially among Americans, is the subject of reams of research and speculation. Two reasons, of course, are manufactured doubt and the building of economic infrastructures around the fossil-fuel industry. Beyond that, the Canadian psychologist Robert Gifford identifies what he calls the "Seven Dragons of Inaction."[5] Although he concedes that climate-averse infrastructures are part of the problem, he sees psychological barriers as a central cause of indifference. Some of these barriers consist of what he calls "discredence" toward experts and authorities, whereby people distrust government officials and scientists, which leads to inaction.

Whether you can afford to be indifferent has in large part to do with how directly you are experiencing the impact of the climate crisis. Consider Yale survey results broken down by race: Hispanics/Latino (69 percent) and African Americans (57 percent) are more likely to be alarmed or concerned about global warming than whites (49 percent). Whites are most likely to be doubtful or dismissive (27 percent) than are Hispanics/Latino (11 percent) and African Americans (12 percent).[6]

People of color in the United States are likely more concerned about climate change because, compared to white Americans, they suffer more often and more severely from environmental hazards, including extreme weather events, sea-level rise, and exposure to air pollution.[7] This disparity holds true for groups that are subject to socioeconomic inequities the world over, including Black and Brown communities in Europe and Indigenous communities everywhere. Throughout the world, the countries with the smallest carbon footprint bear the brunt of the harshest climate impacts, while the most polluting nations—in particular the United States, which tops the list and is responsible for 20 percent of total global carbon emissions, and China a distant second at 11 percent—sidestep responsibility. Widespread evidence of these disparities and growing recognition of the ways structural racism, sexism, and economic inequities exacerbate climate-change effects have highlighted the interconnected nature of injustice and situated environmental concerns squarely in the realm of the broader issues of justice. Indeed, as the Black Lives Matter movement's actions peaked in the summer of 2020, so too did the evidence of the impacts of the climate crisis on people of color and its overlap with structural racism. As the consequences of climate change become more devastating and widespread, a global climate-justice movement—or, more accurately, a movement made up of many movements—has flourished, which sees climate justice through a racial, economic, and intergenerational lens.[8]

Whether you can afford to be indifferent to the climate crisis also has in large part to do with how likely it is to seriously affect your future. A Pew survey in May 2021 on Americans' attitudes about climate change found attitude differences across generations. In response to the statement "Climate should be top priority to ensure a sustainable planet for future generations," a greater majority of Gen Z (67 percent) and Millennials (71 percent) agreed than did Gen X (63 percent) or Boomers and older people (57 percent).[9] Indeed, youth, especially those who are younger than the voting age or who live in authoritarian political environments, are at a political disadvantage because they are excluded from traditional politics. They also have the most at stake. Young people, especially those who live in areas hard hit by climate-crisis impacts, are more vulnerable to the effect of climate change, which has immediate and lifelong consequences for their physical and mental health.[10]

So what should we make of this growing lack of indifference? Writing in the 1920s, both Walter Lippmann and John Dewey contended that when complex and unfamiliar issues emerge, democratic politics are not rendered dysfunctional but rather are enabled. In *The Phantom Public* (1927), Lippmann elaborated this point: "Where the facts are most obscure, where precedents are lacking, where novelty and confusion pervade everything, the public in all its unfitness is compelled to make its most important decisions. The hardest problems are problems which institutions cannot handle. They are the public's problems."[11] This is true, he argued, because straightforward and manageable problems can be addressed through existing institutions and by the communities that they impact. But more complex problems that institutions fail to deal with require publics. In other words, complex issues actually enable public involvement in politics.[12] As Noortje Marres suggests, "Instead of worrying that the complicated issues of today make democracy impossible, we should try to figure out by what amazing means today's issues may bring out the passions of the public."[13]

Our current institutions—media, economic, political, educational—have floundered, not taking the necessary steps to mitigate the climate crisis in large part because they are built on foundations that benefit from maintaining the status quo.[14] But publics have previously spurred foundational change and are poised to do it again. A path to resistance and reform was laid long ago, and it continues to be delineated in the context of contemporary challenges and opportunities. But to clear the path now we need to acknowledge and address the obstacles that litter it.

One major obstacle, as I have been arguing throughout the book, is composed of our media environment and information infrastructures. To achieve social justice, we must first have in place social arrangements that make it possible for everyone to participate on equal footing. That is, equity can't be achieved if everyone whose interests are at stake don't have a seat at and an equal ability to contribute to the negotiating table. The social and political theorist Nancy Fraser calls such an arrangement *participatory parity* and argues that "to overcome injustice we must dismantle institutional obstacles that prevent some people from participating in an equal way with others, as partners in social interaction."[15] Anna Roosvall and Matt Tegelberg point out that patterns that support or obstruct parity in our highly mediated societies are reinforced through media.[16] In other words, the media environment is the negotiating table. Lack of parity there precludes any possibility of parity in other realms—for example, in our physical environment. Hence, struggling for social justice today necessarily involves struggles for media justice. The latter is a path to the former.

This chapter explores what happens after peak indifference, focusing on activists in two mass networked movements, Fridays for Future and No Dakota Access Pipeline (No DAPL), with an emphasis on the ways that struggles for climate justice are intertwined with media and data justice and injustice. These two movements, led by youth and Indigenous groups, respectively, treat the climate disaster as fundamentally a matter of social justice rather than of individual rights, environmental protection, or conservation, which have been central concerns of past environmental movements. They share a commitment to upending the system of injustice and to calling out the willful greed that fuels climate injustice and resistance to change. A close look at how both movements leverage legacy and online media, what obstacles they face, and how they respond to those obstacles offers an illustration of how the struggle for climate justice is integrally tied to the struggle for media and data justice.

Despite the enormous amounts of time, effort, and savvy that activists invest in using media to convey their message, create connections, build solidarity, and mobilize and protect themselves against both online and offline attacks, the lack of parity in their interactions with both networked and legacy media environments can and often do undermine their efforts. Thus, for activists, both legacy and social media act as a double-edged sword whereby media not only are an integral and beneficial part of their movement's strategy but also at the same time present significant challenges and vulnerabilities. Just as climate journalists, no matter how well they do their work, struggle on an uneven playing field, activists, too, face noise and incivility as well as an ambivalent landscape that prizes attention and connection without regard for their social impact.

MEDIA AND DATA JUSTICE

Many of the media strategies of today's climate-justice movements are inspired by those used by the U.S. civil rights movement. In particular, the strategies of combining movement-based media and winning mainstream coverage are often replicated by contemporary environmental- and climate-justice activists. The members of the climate-justice-focused Sunrise Movement, for example, in 2018 staged a sit-in at U.S. Speaker of the House Nancy Pelosi's office to demand that Pelosi work to build consensus in the House for Green New Deal legislation. More than 250 young protesters showed up to occupy Pelosi's office and were joined by Representative Alexandria Ocasio-Cortez of New York. Fifty-one arrests were made during the protest, and the action received massive amounts of media attention as movement-produced videos from the event flooded the internet, which in turn led to mainstream news outlets picking up the story and extending its reach. Cofounder Varshini Prakash says the Sunrise Movement leverages a combination of viral movements to bring people into the movement and the "brass tacks day-to-day grassroots organizing"[17] to give them a way to participate. The movement's strategy, in this case, was very much focused on generating symbolic power, creating a media event to bring attention to the need for more effective climate policy.

Today, though, it is overwhelmingly evident that media wield not only symbolic power but also material power; they shape not only representations but also, perhaps more importantly, the kind of relations forged in media space. In their collected volume *Social Media Materialities and Protest* (2019), the editors Mette Mortenson, Christina Neumayer, and Thomas Poell explain that countless media technologies are employed for the purposes of political dissent, and these technologies always to some extent shape the user activity.[18] Platforms such as Twitter and Facebook, for example, make it easy to mobilize, but they also tend to pull activist efforts away from the arduous work of face-to-face meetings and live coordination, which are still often necessary to create effective long-term strategies.[19] Pulled in this direction, activists focus their efforts on meeting social media demand: they produce instantaneous flows of communication to compete with the constant noise and the co-optation of their political messages; they invest in strategies to avoid or withstand online harassment and other forms of incivility; they create work-arounds to infrastructures that leave them vulnerable to privacy violations; they anticipate and often cater to the way algorithms privilege spectacular images and content over more sober content that addresses wider social and political issues.[20]

In addition to media technologies, data, too, are central to both the symbolic and the material elements of the media environment. The concept of data justice originally referred to collaboration between antisurveillance and social justice activism, driving the former to articulate broader concerns of rights and freedoms and the latter to engage with the technical dimensions of surveillance and resistance.[21] Data justice is taking on new meaning as various forms of data injustice become apparent and as a burgeoning field of data activism sees massive data collection as both a challenge to our rights and a novel set of opportunities for social change. Social media data can tip off law enforcement about when protests will occur and who will be protesting, yet it can also be analyzed by movement organizers to gauge public sentiment and to tailor outreach and mobilizations accordingly.[22]

The data-justice movement emerges from hacker and open-source movements that historically sought to appropriate technological innovation for political purposes. Stefania Milan argues that this new frontier of media activism works to take massive data collection out of the exclusive hands of corporations and governments. It can be reactive, she explains, when activists attempt to resist mass data collection by governments or corporation, or it can be proactive when activists leverage big data in the service of their aims.[23] Reactive activism includes activists using burner phones, secure email and messaging apps, protected virtual private networks, and other tools and practices to ensure that their data are not being traced and used against them. These precautionary measures are increasingly important for environmental activists who are being criminalized in countries around the world, including the United Kingdom and the United States.[24] Proactive data-justice activism includes activists appropriating algorithmic power to spread alternative narratives and advocate for social change. For example, as elaborated later in this chapter, the Fridays for Future and No DAPL movements use centralized hashtags, leveraging their knowledge of algorithms to maximize their visibility, drive the protest narratives, and infiltrate the mainstream-media agenda. Activists are proactively gathering data analytics—through the data behind hashtags, likes, petitions, grassroots funding sites, and so on—to listen to publics and thus better reach them. They are also collecting and using data to monitor or surveille human and environmental abuses. For example, the citizen science project InfoAmazonia strives to capture all pertinent information around human rights, climate, and environmental issues in the Amazon region and so installs and collects data from sensors, scrutinizes satellite imagery, and crowdsources data. It is one among many initiatives whereby members of different publics, sometimes in collaboration with journalists and/or scientists, collect data to identify everything from cancer clusters to deterioration of air and water quality.[25]

Efforts toward both media justice and data justice are a central part of the fight for climate justice. Sometimes this is because activists are passionate about making our information landscape more equitable, and so they consciously and simultaneous work toward climate and media justice. More often it is because of the injustice they face—in the form of surveillance, harassment, invisibility, misrepresentation—as they go about their business of trying to save the planet. That is, activists have to fight for media justice if their other efforts are to stand a chance. Bart Cammaerts famously writes that the power of media activists lies in their ability to understand and leverage what he calls "mediation opportunity structures," wherein they draw on political opportunity structures, which are aspects of the political system that challenging groups have to mobilize effectively. Cammaerts argues that activists today use their knowledge of how the media field and communication technologies operate to identify and leverage chances to resist or shape what is communicated and how.[26] If we consider contests for meaning in the media environment as dependent in part at least on skill to produce, navigate, and even shape a networked environment, we see that media and the ability to recognize and influence its symbolic and material power are key arenas of power and key mechanisms for establishing parity where it does not exist.

CLIMATE-JUSTICE ACTIVISM

As mentioned earlier, the climate-justice movement is made up of diverse, overlapping movements and actions (various boycotts, philanthropic efforts, divestment campaigns, and so on).

Fridays for Future and No DAPL are two movements within the broader movement that share common media strategies, including the use of social media to build awareness and solidarity, live coverage to document and witness protests, and media spectacles to garner mainstream media attention. Activists in both movements also confront similar media-related challenges, such as the noise, incivility, and ambivalence outlined in chapter 2. By paying attention to how these movements navigate the material and symbolic conditions of our current media landscape, we can identify the various forms of disparity or injustice they face, how and when activists are able to overcome or circumvent them, and when they cannot.

Fridays for Future

On August 19, 2021, the *New York Times* ran an editorial by Greta Thunberg, Adriana Calderón, Farzana Faruk Jhumu, and Eric

Njuguna, youth climate activists respectively from Sweden, Mexico, Bangladesh, and Kenya, working with the international youth-led Fridays for Future (FFF) movement. The piece highlighted efforts by youth around the world to step in where adults have failed to protect the climate and demanded that world leaders act at the upcoming UN Climate Change Conference in Glasgow. "We are in a crisis of crises," they wrote. "A pollution crisis. A climate crisis. A children's rights crisis. We will not allow the world to look away."[27]

Thunberg and her collaborators have made their way into the center of climate discourse around the world due in large part to successful navigation of the media landscape. FFF (also known as School Strike for Climate, Climate Strike, Youth Strike for Climate) began in 2018, when then fifteen-year-old Thunberg began skipping school to sit outside the Swedish Parliament and demand climate action. Her strike gained international attention and support, inspiring strikes in countries around the world in which students skip school on Fridays to participate in demonstrations aimed to put pressure on policy makers to act on science and take action to address the climate crisis.

This movement is connecting and giving voice to both today's youth and future generations, whose central message is that those in power need to take the facts and warnings of climate-crisis science more seriously and act accordingly.[28] Justice to them means a science-based response to the climate crisis, with an emphasis on making more visible the communities that suffer the most from the effects of climate change, including the Global South and marginalized communities (Black people, Indigenous people, people of color, women, LGBTQIA+ people) anywhere in the world.[29] The movement stresses equivalence between various struggles and agendas, connecting income, racial, and gender injustices to climate injustice.[30]

The movement forms careful alliances but avoids any party or institutional affiliation. Celebrities such as the actors Jane Fonda and Mark Ruffalo and the popstar Billie Eilish as well as brands such as Patagonia support the movement through making financial donations, showing up to protests, and using media to increase its visibility.[31] The movement's moral authority is boosted by its expressions of fear and anxiety—often illustrated by personal experience—which project authenticity and push back against accusations of alarmism by promoting the idea that fear and panic are reasonable and even motivational responses in the current crisis. FFF is global in scope but with a strong local component and loose international coordination. In the United States, organizers coordinate locally, communicating via Slack and WhatsApp, and spread their messages using social media. Advocacy groups such as the Sunrise Movement, 350.org, Earth Uprising, Future Coalition, Earth Guardians, Zero Hour, and Extinction Rebellion, among others, have joined in to help coordinate actions.

FFF has garnered a great deal of legacy and online media attention thanks in part to its savvy media approach, which involves creating newsworthy events and training members on how to effectively engage with reporters. The internet and social media also play a central role in the movement's communication strategies to connect with and mobilize like-minded participants and to manage and shape mainstream-media representation. FFF uses a range of hashtags, including #fridaysforfuture, #schoolstrike, #climatestrike, and #climatestrikeonline, to aggregate and connect various actors and local chapters. Repeated core hashtags also enable supporters to express their solidarity with the movement through tagging, aggregating, and following stories, comments, memes, and other content.[32] These hashtags then travel with the content wherever it is reposted—Twitter, Facebook, YouTube, Instagram, SnapChat, and TikTok. Memes circulated on social media are used in campaigns to promote messages from scientists, to publicize marches and boycotts, and to encourage climate-friendly behavior such as ditching plastic and recycling.[33] Memes also offer individuals a way to engage in climate discourse and to deal with ecoanxiety—the dread that comes with knowing that we are running out of time to fix things.[34] These strategies are so common today as to seem almost a natural outgrowth of networked communication rather than a strategy of algorithmic justice. But, in fact, hashtags and memes manipulate algorithms and the data they direct to cluster and make more visible certain content by anticipating what the algorithms will promote.

The FFF website also effectively knits together various local actions to underscore the movement's global connections, providing up-to-date data on the size and location of strikes around the world, videos about climate science, reasons to strike, information on the effectiveness behind nonviolent action, and a page aimed at connecting journalists with local climate activists. The recurrent spectacle of the strike—crowds of children gathering in protest each Friday—garnered a great deal of media attention from mainstream outlets, as did the young media savvy leaders in countries around the world, most famously Greta Thunberg, especially before the COVID-19 pandemic drove the actions entirely online.

Members have access to media-relations training through local groups; most of this training focuses on how to deal with legacy media. For example, 350.org offers online training and a handbook called *Let's Get in the News—Media Tips and Tricks*, which teaches youth how to perform well in an interview, how to build and maintain relations with the press, and how to create and stick to talking points—preparing them to be strategic when talking with reporters.[35] As Allie Rougeot, a French activist living in Canada, puts it, "I know that their questions are one-way and then I'll, like twist the questions every time, because I love doing that. I'm, like it's live, you can't stop me now, too bad. You should have known." She adds, "I've been getting better at not giving them any cute sentences or things that it would be easy to get out, so that even if they share a tiny sentence of mine it has at least some sort of impact that I would be happy that it's out there, you know?"[36]

In contrast to this media-positive picture of the movement, however, the greater attention received by Thunberg, who is white and from the Global North, over the larger movement or activists from other countries and ethnic backgrounds has been roundly criticized within and outside the movement. The Seattle-based activist Grace Lambert told me: "What she's doing is great, but she's also not the only one. It's important to think about the people of color who've been doing this work for so much longer than any of us have."[37] In a widely covered incident in January 2020, the Associated Press cropped out Vanessa Nakate, a Black activist from Uganda, from a photo of Thunberg, Nakate, and three other young white women activists who were attending the World Economic Forum in Davos, Switzerland. The backlash to this mainstream-media injustice was immediate and powerful. Nakate posted an eleven-minute video on Twitter, saying, "You didn't just erase a photo. You erased a continent. But I am stronger than ever."[38] Messages from fellow activists supporting Nakate flooded the internet, and the Associated Press apologized and replaced the cropped photo with one that included Nakate. To many, the exclusion of Nakate is an extension of enduring colonialism in international media and even in the climate-justice movement itself, where African countries are often ignored or misrepresented.[39] The swift and strong response by activists suggests not only a recognition of this inequity but also the power to work against it by calling out the media for their unacknowledged racism.

The cropped photo incident is not an aberration. While these young leaders are often portrayed as guileless proponents of a better future, they also endure a great deal of incivility—racism, misogyny, and other forms of misrepresentation and intimidation. The extent of and possible responses to these online threats depend in large part on the type of political regime where activists live. Activists within authoritarian regimes may have as their main priority keeping their identity anonymous or not making any overt political speech to avoid government backlash. They may code political speech and otherwise use what Ashley Lee calls "hidden tactics" to exert influence on public issues: engaging with civic and political issues through artistic and cultural lenses, posting seemingly banal images and text with political subtexts understood only by fellow activists, and so on.[40] Activists in democratic countries tend to more explicitly engage in politics but keep certain details about themselves out of the public eye, such as their address or place of worship, and take care to flag and report abusers.

Youth activists also troll back against their aggressors, a famous tactic used by Thunberg. In 2019, Donald Trump belittled Thunberg's efforts on a Twitter post: "So ridiculous. Greta must work on her Anger Management problem, then go to a good old-fashioned movie with a friend! Chill Greta, Chill!" Following the president's tweet, Thunberg updated her Twitter bio to reflect Trump's comments: "A teenager working on her anger management problem. Currently chilling and watching a good old-fashioned movie with a friend." Her response not only demonstrated her savvy ability to troll her trollers but also garnered much attention from mainstream media: *MarieClaire* called her response "brilliant," and *Business Insider* deemed it "a clever clap back." Her response is part of a strategy that has come to be known as "greentrolling." For example, when Shell tweeted a gimmick poll, asking, "What are you willing to change to help reduce emissions? #EnergyDebate," activists piled on. Representative Alexandria Ocasio-Cortez replied in a tweet: "I'm willing to hold you accountable for lying about climate change." Thunberg similarly replied, "I sure am willing to call-out-the-fossil-fuel-companies-for-knowingly-destroying-future-living-conditions -for-countless-generations-for profit-and-then-trying-to-distract-people-and-prevent-real-systemic-change-through-endless greenwash-campaigns."[41]

Although greentrolling is a way to push back against noise and incivility, it does not stop the harassment experienced by Thunberg and other movement leaders. Politicians and columnists have harped on Thunberg's so-called mental disorders, described her as a "hysterical teenager" in need of a "spanking," likened her activism to "medieval witchcraft," and called the FFF a cult. Some say she ought to be "burnt at the stake"; others have circulated images of a sex doll that resembles Thunberg and purportedly "speaks" using recordings of her voice; still others created and distributed a cartoon that appears to depict the activist being sexually assaulted.[42]

Emily Grey Ellis wrote in *Wired* magazine in early 2020, "While these smears are especially troubling in Thunberg's case because of her age, they mirror the kinds of targeted online harassment employed against many people and groups by those who wish to silence them. The behavior is shocking, but not a shock."[43] Indeed, it is not a rhetorical accident that critics of

Thunberg, nearly seventeen at the time Ellis's article was published, refer to her as a "child." This infantilization invariably comes with accusations of emotionality, hysteria, mental disturbance, and an inability to think for herself, insults traditionally used to silence or undermine the authority of women.[44]

While Thunberg more often ignores or trolls back, some activists pursue legal action. In 2021, Luisa Neubauer, a prominent organizer of Germany's FFF movement, won a court case against right-wing populist blogger Akif Pirinçci for a Facebook post that attempted to sexually humiliate her. She also sued Facebook and won, forcing it to turn over Pirinçci's data. In a *Der Spiegel* interview, Neubauer called on tech companies to take more action against hate speech and roundly criticized Mark Zuckerberg for the company's inaction. "When I look at how slowly Mark Zuckerberg has been reacting to criticism of his company for years, then I don't believe that the situation is going to improve on Facebook in the near future," she claimed, before pointing to the impact of hate speech beyond the internet: "We're not talking about a conflict between me and the hater, but about structural misogyny and sexism that are being expressed everywhere and in ever more radical ways."[45]

While the youngest activists are sometimes shielded from harassment by their parents' or other trusted adults' close monitoring of their social media, there is no filter for many of the youth activists in their teens and twenties.[46] The Seattle-based activist Jamie Margolin described the various types of online abuse she has faced at age nineteen. "Yeah, I get hate all the time. I get people sliding into my DMs telling me things. Some people will call me anti-gay slurs and stuff because I'm very openly out. Some people will say weird and rapey things to me." She was also the target of a neo-Nazi Twitter storm in 2019. "They found out that I was Jewish and that didn't make them happy. They were just commenting like hundreds of comments under my tweets about, 'Where is your synagogue located?'"[47]

Because platforms such as Twitter and Instagram are slow or fail altogether to respond to the abuse, other advocates of climate and media justice are calling out accounts that are targeting young activists, using hashtags such as #CreepyDeniers, #ClimateBrawl, and #TeamMuskOx—referring to the musk ox practice of forming a circle around vulnerable herd members—and engaging in algorithmic activism by clustering the incidents of harassment together to make them more visible.[48] Margolin, together with other activists and mentors, used a WhatsApp group chat to strategize how to collectively flag offensive tweets and report them in order to more effectively bring the harassment to the attention of Twitter. Some are careful to protect their privacy; some avoid posting their faces, locations, and other personal information online. Julia Barnett, who was twelve years old when I spoke with her, said:

> My Instagram is very . . . it is anonymous, sort of anonymous. It has my name Julia Barnett, but my Instagram handle is not my name. And then, I try not to do like full political views or anything on that, although I will post what I'm up to. I don't do like face. . . . I might have accidentally done a brief face reveal once, but really I use Instagram for staying up with what other people are doing for opportunities to apply for climate jobs, and for what's happening with marches and stuff.[49]

Many young activists note that agism and misogyny are not limited to online harassers. Grace Lambert said, "When we're communicating with the city, they always are like, 'Oh, they're like 16.' You kind of get written off as like, 'Oh, they can't make this happen. They can't do this.' I think we're continually having to prove ourselves, which makes it hard." She is incredulous of the ignorance of this take on youth activists: "So many movements have been led by young people, and I'm like, 'Dude, have you not looked at history before?'" Jamie Margolin concurs. "The thing that has really disturbed me," she says, "is the way that it's talked—in the media and by officials—about as, 'Look at what these kids are doing, that's so cute.' There's no follow-through, there's no keeping of your word, there's no real action; it's just, 'Oh, that's cool, the kids march.' But we march for a purpose, and we told you exactly what we need to do, so can you please take action? Then no action actually happens."

Here we can see that young activists are facing media injustice not exclusively in the online environment but also from traditional news media. They are forced to contend with incivility from journalists and political commentators and leaders who won't take them seriously, openly ridicule them, and are intent on applying particular frames to and making invisible (literally in the case of Vanessa Nakate) activists from the Global South. While young people have developed various practices to generate legacy and online media buzz, they have also had to learn to navigate danger and uncertainty in contentious politics and dissent both online and offline by reporting harassment and hate speech, guarding their privacy, pushing back their aggressors through "trolling back" or filing lawsuits, learning not to give the "cute" quote, and using other methods to combat sexism, racism, and ageism. But for many activists, their most effective defense is their ability to influence the narrative both through mainstream media and through their own websites and social media channels. That is, they challenge those who seek to undermine their credibility by creating strong narratives that are personalizing (i.e., the climate crisis is going to affect *me* as a young person who is inheriting this mess) and are informed by moral outrage (i.e., the Boomers did this, allowed this, didn't do enough).

Pipeline Resistance

In June 2021, TC Energy, the company behind the controversial Keystone XL oil pipeline, terminated the project, which it had suspended in January 2021 when President Joseph Biden revoked its cross-border permit. The termination ended a more than decade-long battle that came to signify the debate over whether fossil fuels should be left in the ground. The Keystone struggle mobilized a diverse and forceful movement in opposition to the construction and extension of oil pipelines, many of which run through sacred Native lands in the United States and Canada and all of which pose threats to water and increase carbon emissions, among other health and safety concerns.

The success of the climate activists in opposing Keystone XL was good news for the No Dakota Access Pipeline resistance movement, better known as No DAPL, which has been battling the construction of the 1,172-mile-long underground oil pipeline in the United States known as the Tar Sands Pipeline since it was announced in 2014. The $3.78 billion pipeline is routed across Lakota Nation lands and waterways that have cultural, spiritual, and environmental significance. If TC Energy pulled out of Keystone XL, maybe Energy Transfer Partners and the other companies funding the Dakota Access Pipeline would also back down.

The DAPL path traverses ancient Standing Rock Sioux burial grounds and runs under several bodies of sacred water, including Lake Oahe and the Missouri and Mississippi Rivers.

In resistance to construction, the Standing Rock Sioux (the name of the Lakota and allied tribes) launched the No DAPL movement online, occupied spaces along the DAPL's path, and sued the U.S. Army Corp of Engineers for issuing a permit for the pipeline.[50]

In the past two decades, according to Dina Gilio-Whitaker, whose book *As Long as Grass Grows* (2019) details the Indigenous fights for environmental justice, environmental movements have united with Indigenous people's movements all over the world. She explains, "Environmentalists recognize that the assaults on the environment committed by relentless corporate 'extractivism' and developments are assaults on the possibility for humans to sustain themselves in the future. They recognize that in some ways, what happened to the Indians is now happening to everybody not in the one percent."[51]

In the beginning, No DAPL was a localized movement. Over time, it gained national attention in part due to its media strategies, which won the support of a wider public. Using multiple hashtags, including #DAPL, #nodapl, #standingrock, #StandwithStandingRock, and #waterislife, the movement's members and their supporters circulated memes, news and information, videos, petitions, and funding campaigns using the same algorithmic resistance tactics as FFF to connect with one another, mobilize support, and gain exposure with larger publics and journalists.

Local groups had been actively protesting the pipeline from its initial announcement, but the protests heated up in April 2016 when opponents to the pipeline set up the Oceti Sakowin Camp, where Native and non-Native people from around the world gathered to take part in demonstrations against the pipeline. The number of protesters grew quickly during the summer of 2016, leading to the creation of two additional camps. At its height, the movement represented the largest gathering of Native American tribes in more than one hundred years.[52] While mainstream news media for the most part ignored the movement, each activist camp launched dedicated accounts on Twitter, Instagram, and Facebook that provided users with a steady flow of updates. Simon Moya-Smith, culture editor for *Indian Country Today* and CNN opinion columnist, told *The FADER* magazine: "We have an unprecedented power to shape the narrative and combat false reports in real-time. It's imperative that we embrace the influence of social media and utilize it as a tool to disseminate our own story and to raise awareness of the continued violations of the rights of indigenous peoples on this continent. We survived their white relatives. We will survive their progeny, too."[53]

In addition to these commercial social networks, Indigenous community-based news outlets such as *Indian Country Today Media Network*, *Indigenous Rising Media*, *Lakota Country Times*, and *Native Sun News Today* provided thorough reporting, which in turn provided content for social media pages that promoted sharing and discussion. "There's a lot of things that white people don't want to talk about," said Tim Giago, owner of the *Native Sun News Today* and founder of the Native American Journalists Association. "We remind them of things they'd rather forget."[54]

One of the biggest assets for the protestors was the ability to harness the power of video, especially live video, which lends itself to promotion by Facebook's algorithms. Drone video footage of protest camps captured the attention of the general public, and as law enforcement escalated its tactics to end the protests, the drones caught footage of police violence against peaceful protestors. In November 2016, law enforcement used rubber bullets, water cannons, and tear gas on demonstrators in below-freezing temperatures and later sicced dogs on them. These scenes were recorded, often using drone cameras, and circulated widely, catapulting the events there into the center of public awareness and debate.[55] Shiyé Bidzííl, who captured drone footage over the course of the protests and runs the Facebook page Drone2bwild, said that "simple drone technology has given us activists and grassroots people the upper hand in surveillance and has allowed me to show the world this movement from a higher perspective."[56] Another video activist, Myron Dewey, who began streaming drone footage of the No DAPL protests in August 2016 on his Facebook page, Digital Smoke Signals, shot footage of the November violence that was picked up by national media outlets.

This footage provided evidence of what was happening—both the police violence and the protestors' restraint. "Numerous water protectors who have been arrested have had their charges dropped based on the footage his drones have taken," wrote *Vice*, "and thousands of people have watched as law enforcement have used military-style tactics to suppress the protesters."[57] In an attempt to reduce their own exposure, local law enforcement convinced the Federal Aviation Administration to set up a Temporary Flight Restriction (ending in December of that year) over a four-mile radius surrounding the Standing Rock protests, which applied only to civilians. It was no wonder authorities were pulling out all the stops against the water protectors. The activists' awareness-raising efforts were not only shedding light on the abuse wielded by local police and contracted security forces but also helping to propel a massive divestment campaign against banks funding the DAPL, which the activists called "Mazaska Talks" (*mazaska* means "money" in Lakota).[58] The campaign successfully convinced a number of cities and organizations to take money out of banks funding the

DAPL, including Davis, Seattle, and San Francisco in California. And it garnered the support of celebrities and brands, who helped enhance the movement's visibility.[59]

The sheer reach of the No DAPL network on Facebook is evident in the number of people who "checked in" to the Standing Rock Reservation page, and their effort to disable local police attempts to use Facebook data to identify and target protestors. More than 1.4 million people on Facebook used geolocation tagging to "check in" to the Standing Rock Reservation page in November 2016.[60] Earlier the same month, the American Civil Liberties Union reported that during the Ferguson and Baltimore Black Lives Matter protests, police had been using Geofeedia, a data company, to track protesters via their social media activity.[61] In addition to scouring Facebook and other social media data, local police and the private global risk-management firm TigerSwan engaged in surveillance of the water protectors in the form of aerial surveillance, radio eavesdropping, and infiltration of the camps and activist circles.[62] This information was in turned used against protestors in an attempt to create infighting. An internal document from TigerSwan leaked to *The Intercept* revealed a key recommendation: "Exploitation of ongoing native versus non-native rifts, and tribal rifts between peaceful and violent elements is critical in our effort to delegitimize the anti-DAPL movement."[63] This strategy aligns with a long tradition of leveraging surveillance to try to stir in-fighting and promote agent provocateurs. In addition, Energy Transfer Partners purchased more than 102 antipipeline URLs—including "energytransfer.sucks" and "stopetpipelines.net"—so that they would not be available to actual protest groups, a standard brand-management practice, according to the company.[64]

Once the violence picked up in November 2016, so did coverage of the protests. Most famously, award-winning broadcast journalist Amy Goodman filmed security forces using dogs and pepper spray on protesters. Authorities issued a warrant for Goodman's arrest and alleged that she participated in a "riot," which generated even more coverage and raised questions about the tactics being used to silence the press. In addition to increased mainstream media coverage in the United States and around the world, the protests were the subject of a sketch on *The Daily Show* with Trevor Noah, and Kristen Wiig wore a "Stand with Standing Rock" T-shirt when she hosted *Saturday Night Live* in 2016. Not all coverage did justice to the movement, but the exposure helped put it on the public's and policy makers' radar.[65]

In addition to physical violence, surveillance, and attempts to disrupt footage and criminalize reporting, the protestors faced online harassment. The #yesdapl campaign, for example, threads together insults and misinformation about the protestors and what they stand for. The environmental news outlet *DeSmog* reported fake accounts using the hashtag #nodapl to promote misinformation about the pipeline, claiming that the antipipeline activism kills jobs, that those protesting the pipeline at the Standing Rock Sioux's encampment use violence, and that the pipeline does not pose a risk to water sources or cross over tribal land.

Political bots and sock puppets (fake accounts) are commonly used to disrupt or to influence online discourse, and they were out in full force in the case of #NoDAPL. Russian trolls and bots meant to promote discord within the American political landscape exploited the hashtag #NoDAPL and targeted U.S. energy policy from 2015 to 2017 through the use of Facebook, Twitter, and Instagram accounts controlled by the Russian company

Internet Research Agency.[66] An investigation by the U.S. House Committee on Science, Space, and Technology found more than nine thousand posts produced by 4,334 Russian accounts that dealt with climate and energy issues.[67] For more than a week in October 2016, for example, hundreds of accounts tweeted the #NoDAPL hashtag every six hours, playing up racial and ethnic tensions associated with the pipeline.[68] Ironically, Energy Transfer Partners continues to claim that the antipipeline movement's success was built on its promotion of misinformation.[69]

In February 2022, the U.S. Supreme Court rejected a case by DAPL operator Energy Transfer Partners to avoid a legally mandated environmental review, a major victory for the No DAPL movement. However, the pipeline will continue to operate while the review is conducted.[70]

As of January 2023, the protests have moved north to Line 3, another pipeline project that routes from Alberta, Canada, to Superior, Wisconsin; they are deploying many of the same tactics and have widespread support, but they also have the looming power of the oil industry and law enforcement to overcome. And so the battle continues. The No DAPL movement as well as the Line 3 movement face many of the same media opportunities and challenges as FFF. They, too, must endure online harassment, noise, surveillance, misrepresentation by journalists, and attempts to block and criminalize reporting. However, they also have benefited from a strong, even if stretched-thin, independent press and the aid of a network of solidarity that has lent visibility and gained supporters. And as in FFF's case, their best defense is their ability to shape their own narrative, wresting it away from governments, corporations, and financial institutions with an interest in seeing those pipelines built and flowing with oil.

ACTIVISTS ALONE CAN'T DO IT

These efforts to fight for climate justice driven by youth and Indigenous movements are perhaps part of what is pushing the broader public past peak indifference. Specifically, in the case of Fridays for the Future, through data, memes, hashtags, and narratives young people and their allies have promoted the movement and knit together smaller local actions into a massive global movement, and through training and practices they have developed skills to generate legacy and online media buzz. They have also pushed back against abuses by monitoring and calling out legacy media misrepresentations, by pushing back or "trolling back" against aggressors, by reporting accounts that target kids, and by filing lawsuits. The No DAPL movement has deployed many of the same strategies but has also faced unique challenges. Its activists, too, have used memes, hashtags, collective action, and narratives to gain solidarity. Although both groups have encountered resistance, tactics used against No DAPL protestors have often been more violent and extreme. While FFF youth have faced online attacks and media misrepresentation, Standing Rock protesters have faced all of that and physical violence, rogue government surveillance, ad hoc legal injunctions, and government and corporate sabotage. This greater abuse is due in part, no doubt, to the long history of abuses against Indigenous people that have largely gone unchallenged but also to the fact that their protests interfere directly with the economic interests of the fossil-fuel industry. No DAPL has greatly benefited from drones that capture video footage and independent media coverage that sets the record straight and is picked up on social media networks and spread widely. And although, like FFF activists, No DAPL protestors have developed a work-around to surveillance, they have also been subject to harassments and disruptions—through the tactics of humans and bots—that they have simply had to endure.

Both movements show an impressive ability to leverage mediated-opportunity structures. That is, they are well versed in how the media and communication technologies operate and deftly identify and leverage chances to resist or shape what is communicated and how it is communicated, thus creating dynamic narratives, resisting pejorative framing, withstanding various forms of harassment, and enduring or avoiding surveillance, all the while deftly deploying the media resources available to them. The lack of parity activists face in their dealings with both networked and legacy media, though, bring media justice to the center of their struggles, whether activists intend it to be or not. It's a double-edged sword: the media landscape not only brings the benefits of exposure and connection offered but also subjects activists to often insurmountable and always draining threats and obstacles.

It seems naive to wonder why the "passions of the publics," as Marres calls them, are met with so much resistance and foul play. Clearly, the fossil-fuel industry and its allies will spare no expense and use any method to shut down public efforts to oppose fossil-fuel expansion, and the media environment can easily be leveraged in the service of their efforts.

In chapter 2, we saw how platform infrastructures lend themselves to noise, incivility, and ambivalence. The experiences of the activists detailed in this chapter illustrate the extent to which these qualities shape movement communication. To survive in the contemporary media environment, you have to overcome a vastly uneven playing field, whereby space is afforded to activism but only barely and where to play the game you have to be willing to subject yourself to multiple indignities and injustices.

Even so, although lack of media justice can and often does undermine activist efforts, FFF and No DAPL show that movements today have something that environmental and climate activists of the past several decades did not have going for them: urgency. Most people no longer need to be convinced that climate change is real. The question is: Are they willing to do anything about it? Extreme weather and the corresponding health impacts, reduced agricultural yields, water scarcity, and increased human migration, among other climate impacts, are shaping the everyday lives of people around the globe. While activists show how to hold power to account, we cannot possibly expect them to carry the charge alone. These movement publics need the public at large to join them. There is mounting pressure on all of us to work on behalf of climate action. Activist efforts are increasingly being complemented by actions taken by a wide range of social actors who recognize, as do the movements detailed in this chapter, that the efforts for climate justice are integrally tied to media justice. The next and final chapter highlights efforts by media professionals, lawmakers, and other advocates to remake our media environment so that it can foster rather than preclude effective response to the climate crisis.

4

COLLECTIVE IMAGINARY

In their work of science-based fiction *The Collapse of Western Civilization* (2014), Naomi Oreskes and Erik M. Conway imagine a world devastated by climate change. In their fictional future, the elites of the so-called advanced industrial societies ignored the warning signs of climate catastrophe for decades, mostly just accepting warming temperatures, rising sea levels, widespread drought, and eventually the "Great Collapse of 2393," when the melting of the West Antarctica Ice Sheet caused mass migration and a complete rearranging of the global order. The authors write, "The most startling aspect of the story is just how much they knew, and how unable they were to act upon what they knew. Knowledge did not translate to power."[1]

In the Oreskes and Conway scenario, the road to catastrophe is built on two inhibiting ideologies: positivism and market fundamentalism. Positivism holds that through experience, observation, and experiment, humans gather reliable knowledge about the natural world and use that knowledge to win the power to shape their destinies. But *The Collapse* narrates from the future, where scientists don't hold that power, and whatever power they did hold was surpassed by the power of market fundamentalism—the idea that societal needs are best served by a free-market economic system lightly regulated by government and prioritizing personal freedom. The two ideologies clashed when environmental science made it clear that only substantial government intervention could ward off disaster and safeguard public health and safety. That's when science became the enemy of what Oreskes and Conway call "the carbon industrial complex," and the market overtook science, precluding any possibility of knowledge translating to power.

This dichotomy between markets and science as well as the exclusive blame it lays at the feet of capitalism oversimplifies our predicament, of course. Market tides are turning away from fossil fuel.[2] Moreover, the structures and practices of Western science have at times interfered with effective climate communication and action at least as much as those of journalism and the market have. As in journalists' experience, the commitment to professional practices meant to show the legitimacy of their work has made scientists reluctant to offer predictions or advocate for solutions to the problems unveiled through their work. Donna Haraway, among others, warned decades ago that the scientific norm of neutrality functions as a way to situate knowledge claims not as opinion but as fact and is rooted in masculinity's power to absolve itself of any association or blame.[3] Western science produces knowledge without any recognition that all knowledge production—the questions asked, the methods use, the data collected and interpreted, for what and whose purpose—is always political.

Despite this oversimplification, *The Collapse* offers a sharp description of what does accurately resemble today's contest between the scientific community and the fossil-fuel industry—the tobacco-industry-style fake science and elaborate messaging campaigns paid for with oil company money, the proliferating corporate greenwashing efforts, the deep-pocketed fossil-fuel lobbying groups that stalk state capitols, the Koch brother–funded smear campaigns targeting legitimate climate scientists,[4] the ginned-up passion with which critics, including elected officials, harass teenage climate activists.[5]

Major seats of power view scientific knowledge about climate change as a threat and its champions as enemies and have turned the environmental crisis into an informational crisis that pits the interests of power against the interests of the public. But we have yet to arrive at the Great Collapse—and there are signs that we might yet succeed at avoiding the scenario described in Oreskes and Conway's novel, that we might yet successfully translate knowledge into the kind of power that prioritizes public health, safety, and well-being by rethinking the social contract we have with big tech and journalism in a way that would help repair the epistemological rift characterized by a lack of agreement about what constitutes truth and what sorts of institutions and sources are trustworthy.

Consider two recent congressional hearings that strongly challenged the position that corporations should be trusted to voluntarily work in the public interest.

In October 2021, oil industry executives were called to testify before the U.S. House Committee on Oversight and Reform, which was investigating how oil companies misled the public about fossil fuel's role in the climate crisis and how the industry undermined government action on climate change. In particular, the committee wanted to address fossil-fuel industry advertising that vastly overstates its investments in low-carbon technologies.[6] Committee members drew direct parallels between what the oil industry was doing and what the tobacco industry did. Representative Ro Khanna, a California Democrat and chair of the Environmental Subcommittee, made explicit reference to the infamous performance of tobacco executives at the House "Waxman hearings" nearly three decades earlier: "In 1994, the CEOs of the seven largest tobacco companies appeared before this committee. They too faced a choice. They chose to lie under oath, claiming they didn't know nicotine was addictive. It didn't turn out well for them. I hope Big Oil today will not follow the same playbook as Big Tobacco. You are powerful leaders at the top of the corporate world at a crucial turning point for our planet. Be better. Spare us the spin today, please."[7]

When asked why their companies had denied climate change for decades and were misleading the public by funding misinformation campaigns, the CEOs of ExxonMobil, BP America, Chevron, and Shell were evasive. ExxonMobil CEO Darren Woods said, "As science has evolved and developed, our understanding has evolved and developed." They admitted that climate change was real and claimed that their companies were taking steps toward renewable and clean energy. But when lawmakers asked them to pledge support for ending attack ads on a methane tax and electric cars and to stop lobbying against policy that would reduce emissions, they side-stepped, refusing to make any such commitment.[8]

Some of the driving force behind this investigation came from advertising and public-relations industry professionals who are increasingly mobilized by the climate crisis to fight their businesses' role in supporting industries that pollute both the natural environment and the information environment. In 2015, Christine Arena, executive vice president at the U.S. public-relations and marketing firm Edelman, and five of her colleagues quit the agency because they could no longer stomach supporting its work with fossil-fuel companies and industry front groups.[9] (Edelman represents industry groups such as the American Petroleum Institute as well as corporations such as Chevron and

Shell.) Arena later told *Fast Company* magazine, "I resigned because it became clear that my agency role and professional goals as a climate communicator were incongruous with the agency's business priorities."[10] Leveraging this sentiment, in 2020 CleanCreatives, an initiative launched by the nonprofit media group Fossil Free Media, began a campaign that asked creative companies to sign a pledge to refuse to take on clients who work with agencies who retain fossil-fuel industry clients.[11]

In October 2021, the same month that the oil industry executives testified before Congress, Frances Haugen, the former Facebook data engineer turned whistleblower, testified before the Senate Commerce Committee that the company deliberately hid from the public and government officials its own internal research that documented harms it caused. In particular, she revealed how Facebook's content-promoting technologies, among other things, steered teenagers toward content that promotes anorexia and fuels ethnic violence. The evidence Haugen presented supported what many analysts had been warning about for years. She exposed how the Facebook business model has been and continues to be driven by profits even in the face of solid internal and external research that the platform is currently fueling hate, illness, and distrust in facts and science and thus doing real damage to individuals, societies, and democracy. She argued that allowing Facebook to continue to do business as usual would only fuel "more division, more harm, more lies, more threats and more combat."[12]

Before Haugen left Facebook, she leaked thousands of pages of confidential documents to lawmakers, regulators, and the *Wall Street Journal*, which published a series of reports called "The Facebook Files." Haugen's leaks, testimony, and media appearances put the dangers posed by social media and AI before publics that extend beyond academics, tech industry professionals, and policy makers who have been debating these issues for the past decade at least. She argued powerfully that it matters fundamentally who is creating and controlling the technology and what values those people are prioritizing.[13]

Haugen's exposure of Facebook is one among many recent examples of whistleblowing and other sorts of activism emerging from within the tech industry. A year earlier, the researcher Timnit Gebru was fired from Google after she refused to retract a coauthored paper about how AI language-model software is biased against women and ethnic minorities and about Google's explosive rise in energy consumption and carbon footprint since 2017. Her internal research turned into whistleblowing after she was fired and her research became news.[14] In 2019, Amazon workers led a cross-tech-industry walkout in solidarity with a September climate strike led by youth activists around the world. Emily Cunningham and Maren Costa were later fired from Amazon for organizing a climate-justice group to put pressure on the company to reduce its impact on the climate.[15] And over the past few years, groups such as Climateaction.tech and Tech Workers Coalition, have been established by tech workers to encourage their colleagues across the industry to address the climate crisis by not partnering with oil and gas, reducing the environmental harm, and adopting more inclusive and equitable labor practices.[16]

Taken together, these hearings investigating the dubious practices of the fossil-fuel and tech industries, spurred in part by professional advocacy and in part by political will, suggest we have reached a new chapter in the story of the climate-information crisis, no matter how unwilling the industries on trial are to acknowledge their role in it. The fact that the hearings took place and spurred a torrent of news stories, commentary, and discussion suggests that a wider awareness is growing about the simultaneous unsustainability of fossil-fuel-driven economies and the manipulative toxicity of the information environment created by corporate media outlets and media-technology companies. That these efforts were informed by professionals in the tech and advertising fields and organized by elected officials suggests there is momentum beyond the realm of social movement mobilization.

In a broader sense, the hearings suggest an expanding recognition that when addressing both the climate crisis and the information crisis, we must move away from what Amitav Ghosh calls the "individualizing imaginary in which we are trapped." He writes: "Climate change is often described as a 'wicked problem.' One of its wickedest aspects is that it may require us to abandon some of our most treasured ideas about political virtue: for example, 'be the change you want to see.' What we need instead is to find a way out of the individualizing imaginary in which we are trapped."[17]

Instead of protecting individual and corporate freedoms from regulation, we need to think of the freedom for the human and nonhuman collective to enjoy the benefits of livable physical and informational environments. And instead of just relying on individual-level solutions to ward off the climate crisis— learning to spot disinformation, using parental controls to ward against information pollution, going vegan, or taking public transportation—we need to implement collective solutions that begin with holding polluting industries accountable.

This final chapter considers what it would take to transform our media environment in a way that addresses unequal power dynamics in our information landscapes and creates the conditions by which we can more effectively and fairly engage in climate-related issues. It is, of course, much easier to diagnose the problem than to create solutions, but we need to imagine and articulate possible solutions if we are ever to make them happen. In the spirit of change and the urgency of both the climate crisis and the information crisis, the rest of this chapter discusses steps we can take to hold tech platforms accountable and ways we can transform journalism from the inside. This is not a reinvention of the wheel, nor is it a comprehensive plan but rather a reemphasis of what to me are the most viable paths forward, drawing on my own work and that of other scholars and public advocates.

The jolt of current world events tied inextricably to our information networks—including the COVID-19 pandemic, the reactionary antidemocratic populist political movement in Western democracies, the Trump presidency and its ongoing influence over U.S. politics, Brexit, the Russian invasion of Ukraine—makes it clear that the information crisis must be addressed, that our media environment must be cleaned up. The time is ripe for bold initiatives across the spectrum. Thus, I focus here on how platforms and journalism can be transformed to better support media justice and, in turn, sharpen our ability to respond not only to the climate crisis but also to a host of other current and future threats to livable societies. Ultimately, though, our ability to transform the media landscape will depend in large part on our ability to rethink liberalism, which has had and continues to have a profound impact on our information and natural environments. Indeed, the struggle for climate justice is only the most recent in a long history of challenges to the idea that the freedom to pursue individual interests should be protected above all else. Continuing to recast our collective notion of freedom will go a long way toward cleaning up our physical and information environments.

TRANSFORMING THE MEDIA LANDSCAPE

Transforming the media landscape begins by acknowledging and addressing power imbalances.[18] It is easy to place blame on technology—the problems of platform bias toward extreme content, microtargeting, and bad actors spreading disinformation, among others. The larger problem, however, is tied to media systems, old and new, that have been built to privilege profit over publics. The significant power imbalance between corporations and publics hobble journalism, undercutting journalists' willingness and ability to confront corruption and injustice because in an environment where profit takes precedent, journalists are prone to cater to the very forces they should be holding to account.

Although the work of many individual journalists and outlets exemplifies fairness and equity, some of it detailed in chapters 1 and 3, the default behavior of the commercial legacy news media is to legitimize the inequity on which our society functions. Indeed, journalists are often part of the systems of inequity, as Natalie Fenton and her coauthors explain in *The Media Manifesto*: "It's not about failing to hold banks to account but about the complicity of financial journalists and commentators in celebrating neoliberal economics ahead of the 2008 crash; it's not about failing to be tough on racism but about the media's historic perpetuation of racist stereotypes and promotion of anti-immigrant frames; it's not about failing to recognize the challenges of apocalyptic climate change but about repeating tropes about 'natural' disasters such as hurricanes, heatwaves and forest fires, together with routine 'balanced' debates between climate change scientists and deniers."[19]

The U.S. press has sought to negotiate tensions between profit and public good since the 1920s, when it first commercialized.[20] Public discussion and oversite of journalism and telecommunication throughout the century suggests that there was at least some political will to find collective solutions to address the tensions. In 1943, the Hutchins Commission convened a group of elite media owners and intellectuals to address the view that the press was protecting "the class interests of the moneyed and powerful."[21] Their nonbinding but much discussed report was delivered four years later and declared that the press plays an important role in the development and stability of modern society and that, as such, it is imperative that a commitment of social responsibility be imposed on mass media. The Fairness Doctrine of 1949 mandated that broadcasters cover issues of public importance in a way that represents opposing views. In the latter half of the twentieth century, other protections were put in place to ensure against the concentration of ownership and the preservation of independent media, including antimonopoly laws. But ubiquitous adoption of the internet and other digital tech corresponded with a trend toward deregulation. While the concentration of journalism was already taking place, the Telecommunications Act of 1996 stripped away the government controls that had shaped the media industry since the Communications Act of 1934, which was established to regulate U.S. telephone, telegraph, television, and radio communications. Created to permit fewer but larger corporations to operate more media operations within a sector and to expand across media sectors through relaxation of cross-ownership rules, the new act spurred massive consolidation of media in the United States and paved the way for today's tech oligarchies.

Today, it is darkly comic to think that our profit-driven platformed news and information system was built to operate in any way that prioritizes acting as a pillar of democracy. There is little evidence that the people in charge of platforms or of journalism institutions know what such a pillar would look like or care to pursue it as any kind of priority. These institutions' profits from fossil-fuel ads, their failure to treat the impacts of climate change as a crisis, and their role, even if inadvertent, in spreading disinformation underscore this assertion. The point is that we need a different sort of media system to support a different sort of world that addresses head-on the inequities inherent in both legacy and networked tech-driven media systems.

In the case of both legacy and internet-based media companies, the power imbalance is a result of companies extracting massive amounts of value out of the public in the form of attention, money, and data. At the same time, they are creating the conditions for a great deal of harm to the public, including exposure to bad actors, distorted views of reality, data-industry abuse and surveillance, as well as the representational harms of invisibility and misrepresentation. Simply put, media companies are taking much more than they're giving, and they enjoy liberties such as unlimited growth and autonomy that should be—but currently aren't—conditional on specific commitments to the public good because their role in supporting that public interest is too important to leave to the market.[22]

Our social contract with media companies must be renegotiated. This new social contract and other proposed solutions to addressing the unaccountable power of our media systems have been the subject of reams of scholarly, policy, and popular literature.[23] They are also being worked out in real time as grassroots and professional sectors take on the role of change makers.

TOWARD PLATFORM JUSTICE

Although technology developers, digital rights groups, and surveillance experts have long been concerned with mass corporate and government online surveillance, it was not until the whistleblower Edward Snowden exposed the activities of U.S. and U.K. intelligence agencies in 2013 that pervasive collection of personal data became a global issue of public concern.[24] The Cambridge Analytica scandal in 2018, which revealed that the company had harvested the personal Facebook data of 87 million users to manipulate how they would vote in the U.S. presidential election of 2016, marked another high point of attention around the issue.[25] The story revealed the ways data-collection and sorting practices have transformed election campaigns and other political communication strategies through the use of psychometric data to microtarget advertisements, which amount to covert and deceitful messaging, including massive efforts to dissuade people from voting.[26] It also highlighted the fact that data justice is not just about individual privacy and protection of personal data but also about the power imbalances inherent in datafication and the consequences it has on civic participation and democracy.

Safeguarding Personal Data

In *Surveillance Capitalism* (2019), Shoshana Zuboff argues that safeguarding personal data should be treated as a human right and that regulation should target surveillance economics—the extraction of private life data as an industry asset—for comprehensive reform. Fixes that aim to curb data extraction and its upstream harms are content neutral and thus don't threaten freedom of expression. And these fixes work toward eliminating downstream social harms—toxic content, manipulative advertising, and discriminatory AI—by banning the foundations of the economic practices that enable them. If we get rid of secret extraction, we get rid of illegitimate accumulation of knowledge about people and the corresponding targeting algorithms. Without the ability to target particular groups of people, big tech cannot control information flows and shape people's behaviors to benefit big tech's own interests. We need to regulate extraction in order eliminate profit made from surveillance and in turn the incentive for it.[27] Put in terms of the climate crisis, if we stop data extraction, we eliminate much of the potential profit from climate-change-denialist advertising, and so Facebook and other platforms would no longer be incentivized to do business with these groups.

Governments are now beginning to take steps to rein in data extraction. The European Union has written the General Data Protection Regulation (GDPR),[28] which gives people the right to demand access to their personal information and to demand that organizations destroy any personal information they may have compiled. The GDPR has become a model for laws being written around the world, including in Turkey, Mauritius, Chile, Japan, Brazil, South Korea, Argentina, and Kenya. Despite no longer being an EU member state, the United Kingdom retains the law in identical form. The California Consumer Privacy Act, adopted on June 28, 2018, has many similarities with the GDPR. There are shortcomings to this approach, though, most notably that it focuses on individual data protections rather than on recognizing and finding solutions to the collective problem of mass extraction.[29] But it is a start.

Many argue that the most effective approach to limiting abuses is to center efforts on breaking up the tech oligopoly. Tim

Wu makes the case this way: "From a political perspective, we have recklessly chosen to tolerate global monopolies and oligopolies in finance, media, airlines, telecommunications and elsewhere, to say nothing of the growing size and power of the major technology platforms. In doing so, we have cast aside the safeguards that were supposed to protect democracy against a dangerous marriage of private and public power."[30] These safeguards include antitrust and competition laws aimed at reshaping market structure, reducing size, limiting growth, and putting guardrails around market behavior. Antitrust laws can work toward leveling the playing field by limiting private concentration of economic power.[31]

If companies were required to separate their advertising businesses from the communication networks through which people connect with one another, there would be less profit incentive to amplify and spread content based only on whether it generates attention.[32] The separation would break up the dangerous concentrations of power over speech and could also encourage competition that would offer greater privacy protections, content selected for qualities other than being attention getting, and protections against harassment and abuse.[33] As it is now, journalists and members of the public are the only ones guarding against the blurred line between organic and paid content. When in 2020 Google tried to pass off paid content in search results as "organic,"[34] it would have gotten away with the scam if it hadn't been for pushback from journalists and commentors amplified on social media.[35] Google runs ads on websites and YouTube videos that promote climate misinformation, despite pledging to stop doing so in October 2021, so it's safe to assume these ads and videos were among those pushed as "organic content" before Google was called out for doing so.[36]

Holding Platforms Accountable

Other policy advocates focus on curbing content and AI-based harms—that is, on what Zuboff and others call "downstream solutions."[37] Although not addressing the root problem, this approach can offer stopgaps to the rush of problematic content and the algorithms that push it. Some argue that platforms ought to be held accountable for the content they host and the harms it might cause. In the United States, platforms are protected by section 230 of the Communication Decency Act of 1996, which provides internet service providers safe harbors to operate as intermediaries of content without fear of being liable for that content as long as they take reasonable steps to delete or prevent access to that content. Personal liability for executives and engineers would incentivize, as it does with new medicines, the view that new technologies should be evaluated for safety and efficacy before they are released on the market. And it would spur platforms to put more effort into getting rid of climate disinformation, hate speech, and conspiracy theories, all of which affect climate-change discourse, if they no longer have blanket immunity for harm caused by third-party content. Platform executives and engineers would put in place more rigorous and fair content-management practices, and they would stop designing algorithms that push content based on attention and instead simply post content in reverse chronological order, which was the original organizational framework for newsfeeds or other media to treat all content the same.[38] Policy is being developed to address this reform. For example, the European Union Digital Services Act of 2022 requires companies to be responsible for the content on their platforms or face steep fines.[39] Companies must set up new policies and procedures to quickly remove flagged hate speech, terrorist propaganda, and any other content that is illegal in countries within the European Union. Such measures, though, raise concern about giving companies that host third-party content the power to make decisions about whether content violates the law or standards of decency and about whether an organization is a "terrorist" group. Even so, it seems the tide is turning against treating platforms as intermediaries with no responsibility for content. Like mass-media companies before them, platforms need to sort out systems of identifying and getting rid of harmful and illegal content.

Redirect Tech Platforms' Profit to Support Journalism

The most pressing change we need to make in journalism is to decouple it from the market. A growing body of literature on the future of journalism underscores the need for journalism that is not driven by profit. Victor Pickard, for example, argues that journalists and those who study the industry must fix what he calls the "supply-side problems," or reliance on the market for financial stability, in order to solve the "demand-side challenges," namely the public's lack of trust in and thus engagement with news.[40] "In the United States," he writes in his article "The Violence of the Market" (2019), "we treat the market's effects on journalism—as we treat the market's effects on nearly everything—as an inevitable force of nature beyond our control or, at the very least, a public expression of democratic desires. This 'market ontology' simultaneously naturalizes the market's violence against journalism and forecloses on alternative models." The result of this resignation is to preclude any possibility that society will attempt to fix this major social problem through changes in public policy.[41]

We cannot, however, afford to be resigned.

A burgeoning movement seeks to push big tech to fund legitimate journalism. The platforms extract great benefit and profit from users, the thinking goes, so they should have to pay to help alleviate some of the harms their services have facilitated—polluting the news and information systems, abusing their market power over digital advertisers, displacing traditional news media with media that seek only revenue. But major potential drawbacks to big tech funding the news business also need to be considered when planning how this model should be implemented.

The most obvious conflict of interest is that direct funding by big tech would likely inhibit reporting. As Emily Bell, the founding director of the Tow Center for Digital Journalism at the Columbia Journalism School, points out, "Facebook, Apple, and Google do things that journalists should be investigating, not profiting from. . . . All three have strategies for managing the press, and they publish very little data about what happens on their platforms or what the effect of it is, making tech reporting a vital form of accountability."[42]

Bell is commenting on Google's partnership in 2019 with U.S. newspaper giant McClatchy to fund three new local news outlets: the *Telegraph* in Macon, Georgia; the *Modesto Bee* in Central Valley, California; and the *Kansas City Star*. Just a year after the partnership was announced, McClatchy filed for bankruptcy, and all of its papers were taken over by the hedge fund Chatham Asset Management. McClatchy executives blamed its failure on the cost of pensioners, who vastly outnumbered current employees. News industry analysts, though, saw the bankruptcy as the inevitable result of a company expanding by accumulating debt while its base of customers was shrinking.[43] Hence, not only did the McClatchy/Google partnership compromise participating outlets' ability to hold big tech accountable, but it also failed to help the company stave off bankruptcy and hedge fund takeover.

Other, better arrangements could facilitate big tech/journalism partnerships in a way that would less likely compromise journalists' independence. The journalist and lawyer Lucia Walinchus calls for establishing a national nonprofit news utility, for example, where a portion of digital-advertising dollars would go to journalists.[44] This is not a radical idea. Similar steps have been taken before. For example, a portion of U.S. customers' cable bill goes to C-SPAN because C-SPAN helps us keep track of Congress.[45] In the United Kingdom, what's called a TV "license" is a tax that goes to the BBC. A news utility would function in a similar way; whenever you buy digital advertising, part of that money would go to journalists and local newsrooms. Pickard and Bell suggest that money from either big tech or the government be put in an independent trust for journalism and allocated by an independent third party.[46] Both plans would allow for a shift away from large news corporations that consolidate all journalism markets at a national level to new funding models that support local community-based journalism. The climate crisis demands local, community-specific news outlets to engage the public about issues that are quite literally close to home—specific reasons to care whether the issues are logistical (the water is rising) or value related (rising water is flooding farmland and taking away the livelihood of agriculture workers) or both.[47] Local news is preferable, too, because planning for the future impacts of the climate crisis often takes place at the level of local authorities and involves multiple stakeholders. Under these circumstances, local media have a critical role in communicating these differing perspectives within communities.[48] There are various ideas about how to allocate the funds to journalism: Walinchus suggests people collectively decide on standards for obtaining the funds and then vote on it, and Bell advocates for having a third party decide. Whatever the specifics, the point is that big tech could renegotiate their social contract by supporting journalism. The fundamental question then becomes: What kind of journalism do we need?

Reforming Journalism

Putting aside nostalgia for the era when the authority of experts went relatively unquestioned and fact-based evidence was seen as the only essential part of public debate, we need to acknowledge and repair the ways journalism has been and remains a large part of the problem.

There is no going back. The truth is that legacy journalism historically has excluded the vast majority of voices and points of view, mostly ignored any authority rooted in the experience of life in the non-Western world, and dismissed evidence not based solely on rationality but on morality, belief, personal experience, and emotion. We need to take a hard look at the question of whom journalism is failing to serve.[49]

This deep look will require self-reflection by journalists. Currently, there is very little acknowledgment of or engagement with the representational harms of journalism historically misrepresenting and distorting issues related to communities of color, women, labor struggles, and a host of issues—such as the climate crisis and income inequality—that challenge status quo arrangements.[50] In their book *Reckoning: Journalism's Limits and Possibilities* (2019), Candis Callison and Mary Lynn Young call out journalism scholars' role in the lack of reflection, wondering why questions related to gender, race, and colonialism are so often ignored among researchers who study a profession that aims to speak truth to power. They point out that although journalism scholars regularly reference the emergence of objectivity in the early twentieth century, these same scholars rarely acknowledge that during the same period journalists ignored the fact that Indigenous people and people of color in the United States and Canada were suffering myriad grave injustices because they saw such injustices as a normal part of the social order. Indeed, Callison and Young argue that one of the central challenges for journalists and journalism scholars today is to figure out how to narrate the present in a way that acknowledges power relations, exclusions, and the harm wrought by mis- and underrepresentation.[51]

One way to do this work is to expand who is considered a legitimate source of authority and thus include nonwhite people and people with expertise and leadership in communities that are often ignored or disparaged in the news. Journalists gain authority through the sources they rely on, and sources gain visibility as experts when journalists rely on them.[52] Journalists have the power to expand what and who we think of as legitimate sources of authority and so to carve out greater space for equity in public discourse.[53] The people left out of news accounts are the people who have long been marginalized in many other ways as well—largely ethnic minorities and low-income people. Of course, the worst of the climate crisis, as with so many other crises, will disproportionately affect and inflict suffering on marginalized people. It is more important than ever that journalists expand the circles of what communities are represented and who within those communities is considered the elite so that democracy works for more and more diverse types of people. It is easy for people with property, wealth, and power in different parts of the world to continue to downplay environmental degradation and disaster, but the hundreds of millions with little resources crowded onto sinking islands and into continental lowlands, river deltas, hurricane alleys, wildfire corridors, drought-dried subdivisions, refugee camps, and makeshift rafts around the world know what's coming. For them, the climate crisis is already here. Journalism should redouble its efforts to prioritize these mostly disenfranchised groups and thereby make its work more relevant, compelling, urgent, and just.[54] It should provide space for their voices, focus on issues that affect them, acknowledge the expertise of leaders and others in these communities, bring members of these communities into the newsrooms, and create supportive work environments.

These changes would not only offer visibility to sources beyond the elite but also help restore the legitimacy of journalism among the growing number of people who have come to mistrust elites—who they believe, as Barbie Zelizer, C. W. Anderson, and Pablo Boczkowski do, have "cracked up," causing a disintegration of the system of elite governance that journalists relied on to index their information gathering. These changes will also help remedy one of the most vexing problems of journalists today, according to Zelizer, Anderson and Boczkowski: "Journalists are increasingly forced to choose between either representing the range of important political opinions that actually exist or holding fast to their liberal foundations as democratic enablers."[55] In other words, they must choose between representing the range of opinions that actually exist or acting as a megaphone for the political elite and thus drumming up support for their campaigns or policy platforms.

A wider if not entirely representative range of opinions can be found online and in the hybrid journalism-adjacent spaces where journalists, activists, sources, and tech developers/coders are creating open-source networks and a wealth of news-related information, opinion, and cultural expression in different styles and from various producers, all of which are reshaping the meaning of news events and issues.[56] Even before digital tools and networks became ubiquitous, more people were doing acts of journalism than were acknowledged.[57] As we have seen in the case of leading climate journalism profiled in chapter 1, the changing circumstances in which many climate-specific news sites operate have created the conditions for more collaborative relations among journalists and others actively working in the field of producing climate-related communication. In the more optimistic days of 2013, Andrew Chadwick wrote that "hybridity is creating emergent openness and fluidity, as grassroots activist groups and even lone individuals now use newer media to make decisive interventions in the news-making process."[58] Although we have seen the ways hybridity can engender deeply dysfunctional interventions in the news process, it can and has helped expand the definition of a legitimate source of authority to include people with expertise and leadership in communities that are often ignored or disparaged in the news.[59]

In addition to reflecting on and expanding who is included in and served by journalism, journalists need to start advocating for a more equitable media environment, one in which people have the right not to be poisoned or systematically excluded rather than the toxic one built by Silicon Valley. Journalism must push the public to understand what is at stake and pressure the media industry to improve. It's no secret that journalists famously avoid covering the journalism industry because doing so challenges their dedication to neutrality. How can journalists be neutral about the fact that hedge funds own more than half of U.S. newspapers and are buying up and gutting many more? Decades ago, journalists decided, one way or another, that it was better to give only scarce attention to the biggest business and democracy stories in the country.[60] The drive toward a more equitable media environment isn't about journalists putting out more and better information and hoping for the best; it's about journalists pushing themselves to join with other media professionals—the public-relations and advertising professionals refusing to serve fossil-fuel clients, the tech workers demanding better climate policy from their companies—to leverage their power and make the case for a more equitable and healthy information and media landscape.

More reporters are delivering exposés on the nefarious side of big tech. The *New York Times* reports on the way credit card companies collect data on their customers' mental health to predict future financial distress, so that a visit to a marriage counselor might translate into downgraded credit ratings.[61] *ProPublica* reports on Facebook algorithms programmed to deliver housing ads that screen out people of color, women, and older workers; the *Wall Street Journal* reports on the way prices for products or enticements to buy change depending on a shopper's zip codes, paradoxically often hiking up prices for people in hard-pressed neighborhoods; and NPR reports on how climate denialism flourishes on Facebook despite the platform's pledge to ban it.[62] What these stories have in common is they are efforts to inform the public about what these companies are doing and in turn to hold the companies accountable for what otherwise would be largely under-the-table but highly impactful behavior. A steady drumbeat of this kind of reporting is building where just a few years ago very little watchdog reporting took place and tech cheerleading in the media was practically a business model. As Zizi Papacharissi describes this kind of reporting, "Journalists can be democracy's conduits of trust, and, in doing so, be agents of change."[63]

Finally, journalists must help the effort to break down the cultural narratives that hold us captive and that slow action on addressing the major issues of the day. Most fundamentally, we need journalists to help recast the powerful dominant narrative around freedom.

RETHINKING FREEDOM

Western philosophies of freedom have centered primarily on threats to liberty posed by humans. Liberation from constraints imposed by nature were simply understood as a perquisite of humanity. According to the historian and postcolonial theorist Dipesh Chakrabarty, "Nonhuman forces and systems had no place in the calculus of liberty: indeed being independent of nature was considered one of the defining characteristics of freedom itself. Only those who had thrown off the shackles of their environment were thought to be endowed with historical agency."[64] But nature is now increasingly seen as part of the material reality humans must reckon with, on the same spectrum with our machines and technologies and waste.[65] This changing view is about easing categorical thinking and emphasizing continuity. It reflects long-established Indigenous epistemologies that see humans and nonhumans as kin[66] and more recent work in Black ecological thought aimed at developing ways of living with and in nature that break from exploitative traditions of capitalism and settler colonialism.[67]

The political theorist Jane Bennett argues that we need to recognize the active participation of nonhuman forces in constructing our reality; we need to recognize what she calls this "vibrant materiality." Bennett believes this kind of materiality creates greater opportunities for productive, nonapocalyptic visions of the future. It makes no sense, she argues, to withdraw from nature in an attempt to protect nature when the health of the planet is increasingly reliant upon human intervention.[68] Frameworks that imagine humans more fully integrated with the natural world allow us to think about ways humans can be involved in more reciprocal relationships with nature and ways technology can breathe life back into the natural world that it is now mostly destroying. Bennett situates our freedom squarely in relation to nature.

Today, not only are Western mainstream notions of freedom still hitched to the separation between humans and nature, but they're also firmly grounded in the tenets of liberalism that privilege individual over collective freedoms. Whitney Phillips and Ryan Milner argue that the dominant conception of freedom of speech needs to be updated: "It's the difference between asserting that an individual has the right to spew whatever poison they want without restraint and asserting that those within the collective have the right not to be poisoned."[69]

Regarding fossil-fuel pollution, the tensions around freedom boil down to asserting that an individual has the right to live their life as they always have versus, again, asserting the collective right not to be poisoned. This thinking emphasizes the tension between an individual's freedom to use energy and the communal freedom to enjoy a livable world. As Chakrabarty puts it, "Most of our freedoms are energy intensive. Imagining a low-carbon world, then, means reevaluating our conception of freedom itself."[70] For now, of course, it seems easier for many of us to imagine a tattered unlivable planet than to imagine humans intentionally transforming industrial economies, especially at the scale demanded by the climate crisis.

But we *have* done this kind of transformational work in the past. Advocates of climate justice point to the abolitionism of the eighteenth and nineteenth centuries. The psychiatrist and author Robert Jay Lifton reminds us in his book *The Climate Swerve* (2017) that to prevent a two-degree planetary temperature rise, the fossil-fuel industry would need to forego roughly $20 trillion in untapped wealth. "The last time in American history that such extraordinarily valuable stranded assets existed was in 1865, and the 'assets' took the form of human beings," he writes in *Dissent* magazine.[71] Abolitionism required an upending of an entrenched economic system, and so too will a transformation away from a fossil-fuel-based economy. That's not where the link to slavery ends, though.

The journalist and political commentator Chris Hayes points out that wresting control from those who stand in the way of building more environmentally sustainable energy infrastructures will include conflict, just as ending slavery did. Powerful entities, he argues, will never voluntarily surrender their wealth, and they will fear the fact that dismantling one part of the power structure will likely radically alter that structure, pushing them away from the levers of power.[72] This is how climate justice and racial justice are inextricably intertwined and why it is so important to see them as intertwined as we build activist strategies.

It is worth quoting Robert Bullard, the father of environmental justice, at length here to bring home the interrelated nature of climate and racial injustice:

> When certain lands are seen as exploitable, the people that happen to be living there are viewed as expendable. Hence the genocide of Native people and the exploitation of slaves. At the same time, pollution is seen as just a byproduct of moving to the highest level of the economy. Smoke, air pollution, water pollution . . . that is the smell of progress. When it comes to waste, waste will flow along the path of least resistance. Historically, this means

places near populations that have "less value" and have less political clout to resist. And you end up having places that are "sacrifice zones," zones that are viewed as expendable when it comes to the environment, population, and health. When laws and regulations are built with this mindset, you end up assigning a value to those locations, and somehow that is where the waste and other externalities go.[73]

The comparison between abolitionism and climate activism underscores the daunting scale of the work. Yet it also offers hope, given the eventual although not complete success of the abolitionist movement and the understanding of freedom it introduced. There are other precedents as well. Lifton comments: "If climate change means redefining freedom, we can look back to the social movements—not just abolitionism, but feminist, labor, and anti-imperialist struggles—that won the civil and democratic rights we enjoy today. None could have foreseen the extent of the climate crisis, but the paradigm shift they imagined—toward collective values—are [sic] echoed by the writers and activists today who are calling for a broadening of the commons and for viewing ourselves as stewards, not masters, of the environment."[74]

The climate crisis is the latest in a long line of powerful challenges that underscore the inadequacy of the libertarian conception of freedom that prioritizes individual interests. Today's climate-justice advocates are working to change this individualizing imaginary. Youth and Indigenous activists are making the case that legitimate freedom also must include serving the rights of future generations and the rights of nature. They are advocating for the freedom to pursue collective freedoms, to hear a plurality of points of view, and to oppose the fossil-fuel industry. These efforts toward redefining freedom have involved leveraging a revised media landscape. Media justice and struggles for collective freedoms have always gone hand in hand because media have always been an essential tool in mobilizing counterpublics and for these counterpublics in turn to gain access to dominant publics.

Today there is a growing sense of urgency not just among members of the public but also among lawmakers who are passing laws that support both energy transition and media reform. Today there are increasingly legitimate non-carbon-producing and cheaper alternatives to fossil fuels. Just a decade ago, the question was: Can we create sufficient alternatives? Today the question is: Will we decide to adopt them in time? There are also an expanding number and category of leaders—cultural, religious, political, international, national, local—committed to addressing not only the climate crisis but also environmental and climate injustice.

These developments are making a difference. Consider the recent string of significant victories.

The long-running effort on college campuses to force university administrators to divest from fossil-fuel companies has been particularly successful. Harvard University's $42 billion endowment is no longer tied to fossil-fuel-industry profits. The list of schools that have divested includes Boston University, the University of Minnesota, Georgetown University, Cornell University, and the University of California. By 2015, fossil fuel was the target of the fastest-growing divestment movement in history, and by October 2021 a total of 1,485 institutions representing $39.2 trillion in assets worldwide had committed to divest partially or entirely from fossil fuels.[75] Bill McKibben wrote in a *New York Times* piece that month: "Divestment has helped rub much of the shine off what was once the planet's dominant industry. If money talks, $40 trillion makes a lot of noise." He continued: "When we began the campaign, our immediate goal was, as we put it, to 'take away the social license' of Big Oil."⁷⁶

In the meantime, May 2021 saw two major victories against oil industry giants. ExxonMobile and Chevron shareholders turned their discontent regarding the corporations' direction into a boardroom pushback. A small hedge fund, Engine No. 1, successfully replaced two Exxon board members to help force change. It was leveraging the fact that board members were critical of the company's slow adaption to the evolving energy sector, its meager efforts to reduce carbon emissions, and its claims to the contrary in consumer ad campaigns. Soon after this replacement, a majority of Chevron shareholders demanded the company's board vote in favor of an activist proposal to cut carbon emissions.⁷⁷ That same month, the Hague District Court in the Netherlands ruled in favor of climate activists who were demanding that Shell lower its emissions by 45 percent from 2019 levels by 2030, a much faster rate than Shell had planned.⁷⁸

There were also major community-led grassroots victories. A predominantly Black neighborhood in Memphis, Tennessee, successfully halted construction of a forty-nine-mile underground pipeline. Authorities that included the governors of New York, Pennsylvania, Delaware, and New Jersey as well as the North Atlantic division engineer of the U.S. Army Corps of Engineers agreed to ban fracking in fourteen-thousand square miles of the Delaware River Basin. Los Angeles County stopped all new and existing oil- and gas-drilling projects, including in the Inglewood Oil Field, the largest urban oil field in the country, which recently spilled 1,600 gallons of oil, polluting the surrounding predominantly Black neighborhoods.⁷⁹

The Inflation Reduction Act of 2022 includes the first major climate legislation in U.S. history: a commitment of $370 billion to curb emissions and hasten the clean-energy transition. The act puts the United States, by far the world's biggest polluter, on course to reduce emissions by 40 percent from 2005 levels by 2030. It also attends to the needs of low-income people and people of color by including, among many other provisions, funding to pay electric bills; tax credits to make homes more energy efficient; investment in cleaning up pollution and environmental "hot spots" created by racial redlining that have led to disparities in health and other climate-related threats; and resources to address damaging transportation infrastructure and highway projects that have caused displacement, disinvestment, and economic isolation.[80] After decades of inaction, this bill demonstrates there exists enough political will to finally land the climate crisis at top of the U.S. government's agenda, which is in no small part due to the groundswell of activism on the streets, on university campuses, in boardrooms, and in tech and media sectors.

These victories and many more are testament to the way publics are cutting through information pollution. Despite the noise, they are leveraging communication channels to successfully take on, even if incrementally, some of the most powerful corporations the world has ever known, forcing them to make concessions that prioritize collective rights over individual rights.

Much more bold action will be required. This book set out to reframe the climate crisis as both an environmental crisis and an information crisis precisely because we must work in the media environment to change the energy industry and keep carbon in the earth and out of the sky. We have made this mediated world. We live mediated lives. We can't surrender the mediated space.

ACKNOWLEDGMENTS

I am grateful for inspiration and help in various forms that came from all corners all along the way.

First, thanks to scholars whose work helped shape this book and whose friendship sustained me over years of pandemic shut-in and book writing. To Risto Kunelius for talking through every idea and reading every word more than once, thank you! To Mike Ananny, Melissa Aronczyk, Lance Bennett, Bart Cammaerts, Nick Couldry, Stephanie Craft, Amy Flynn, Christine Harold, Dominic Muren, LeiLani Nishime, James Painter, Matt Tegelberg, Karin Wahl-Jorgensen, and Silvio Waisbord for teaching me new things about how to think about publics, technologies, journalism, and the climate crisis. To Zizi Papacharissi, Erika Polson, and Lynn Schofield Clark for all of that, plus Zoom happy hours. To Crystal Hall and Dominic Muren for rainy-porch happy hours. To all of you for the ready reminders that my colleagues are what I like best about the work I do.

Thanks to Matt Powers and Ekin Yasin for their always-lively feedback, for their commitment to communal approaches to work and fun, for repeat lessons in seizing upon good ideas and making the most of fleeting moments. Thanks to Isabel

Münter for always being down to talk big ideas and for fighting the good climate fights.

Thanks to Elizabeth Weber for the cover photo, taken in April 2019 on the Hawaiian Island of Molokaʻi. Traveling with her there to photograph and clean up a small bit of the massive amount of plastic washed ashore and experiencing the energy and creativity she brings to her activism and art were part of the inspiration that brought this book to life.

Thank you to Robert and Anne Russell for their steadfast and much appreciated support.

Thanks to Jay Duchene, who taught me sustainable living from the very start.

Thanks to the University of Washington Department of Communication for providing me with the space and time to do research and to write. To the Mary Laird Wood Endowment for its generous funding. And to Seonah Kim, Jenny Lee, Hai Wang, and Abigail Shew for their meticulous and skilled research assistance. To David Lildawen for the last-minute help. To the MediaClimate research group members, including (again) Risto Kunelius and Elisabeth Eide, from all of whom I've learned so much and with whom I have had so much fun over the years. Thanks to Cardiff University Future of Journalism conference organizers for inviting me in 2019 to talk about the ideas that launched this book. Thanks to Philip Leventhal, Michael Haskell, and Annie Barva at Columbia University Press.

Some material in these pages appeared first in different form as published journal articles and research reports. Interviews with niche news-site founders and editors were originally conducted with Jarkko Kangas, Risto Kunelius, and James Painter for work that appeared in "Niche Climate News Sites and the Changing Context of Covering Catastrophe," *Journalism: Theory, Practice, Criticism*, online, July 6, 2022, https://doi.org/10.1177

/14648849221113119; and "The Journalism in Climate Change Websites: Their Distinct Forms of Specialism, Content, and Role Perceptions," *Journalism Practice*, online, April 30, 2022, https://doi.org/10.1080/17512786.2022.2065338. That research was funded by the Helsingin Sanomat Foundation. Research on science activism was originally conducted with Matt Tegelberg and published in the article "Beyond the Boundaries of Science: Resistance to Misinformation by Citizen Scientists," *Journalism: Theory, Practice, Criticism* 21, no. 3 (2019), https://doi.org/10.1177/1464884919862655. Interviews with young activists were conducted as a part of Mapping Youth Activists and Their Media Experiences project, funded by the Helsingin Sanomat Foundation.

Finally, thanks to the journalists, activists, and tech developers who have been so generous with their time, talking to me about the work they do and why they do it.

And to my favorite journalist, John Tomasic, always interested and full of ideas, beginning to end.

NOTES

INTRODUCTION: TWO CRISES

1. Carson had an eclectic career as a scientist, science writer, and editor, which included regular contributions to journalism outlets. Her first book, *Under the Sea Wind* (1941), was serialized in the *Atlantic Monthly*. *Silent Spring* was serialized in the *New Yorker*, and early in her career she wrote articles on science and nature for the *Baltimore Sun* and other popular mainstream news outlets.
2. *Time* quoted in Mark Stoll, *Rachel Carson's* Silent Spring, *a Book That Changed the World, Environment and Society*, virtual exhibition, 2012, http://www.environmentandsociety.org/exhibitions/silent-spring/overview.
3. Eliza Griswold, "How 'Silent Spring' Ignited the Environmental Movement," *New York Times Magazine*, September 23, 2012, https://www.nytimes.com/2012/09/23/magazine/how-silent-spring-ignited-the-environmental-movement.html.
4. Dan Hallin, "The Passing of the 'High Modernism' of American Journalism Revisited," *Political Communication Report* 16, no. 1 (2006): 14–25.
5. John Hartley, *Understanding the News* (London: Methuen, 1982).
6. Hallin, "The Passing."
7. Ida B. Wells-Barnett, *Selected Works of Ida B. Wells-Barnett*, ed. Henry Louis Gates Jr. (London: Oxford University Press, 1991).
8. Upton Sinclair, *The Jungle: The Uncensored Original Edition*, ed. Earl Lee (Tucson, AZ: See Sharp Press, 2003).

9. IPCC, "Climate Change 2022: Impacts, Adaptation, and Vulnerability," n.d., https://www.ipcc.ch/report/ar6/wg2/. Criticism of the oil and gas industry's "climate-blocking activities" was edited out of the final draft of the 2022 IPCC report according to its authors. See Amy Westervelt, "We Can Tackle Climate Change If Big Oil Gets Out of the Way," *The Guardian*, April 5, 2022, https://www.theguardian.com/environment/2022/apr/05/ipcc-report-scientists-climate-crisis-fossil-fuels.
10. Lelani Nishime and Kim D. Hester Williams, eds., *Racial Ecologies* (Seattle: University of Washington Press, 2018).
11. Robert Bullard et al., "Climate Change and Environmental Justice: A Conversation with Dr. Robert Bullard," *Journal of Critical Thought and Praxis* 5, no. 2 (2016), https://iastatedigitalpress.com/jctp/article/566/galley/446/view/.
12. Natalie Fenton et al., *The Media Manifesto* (Cambridge: Polity, 2020).
13. For an overview of the MediaClimate group, see https://mediaclimate.net/; for results of research on coverage of UN COP meetings, see Risto Kunelius and Elisabeth Eide, "Moment of Hope, Mode of Realism: On the Dynamics of a Transnational Journalistic Field During UN Climate Change Summits," *International Journal of Communication* 6 (2012): 266–85, https://link.gale.com/apps/doc/A287109746/AONE?u=wash_main&sid=AONE&xid=4d20adob ; and Elisabeth Eide and Risto Kunelius, eds., *Media Meets Climate: The Global Challenge for Journalism* (Gothenberg, Sweden: Nordicom, University of Gothenburg, 2012). For results of the IPCC study, see Risto Kunelius, Elisabeth Eide, Matthew Tegelberg, and Dmitry Yagodin, eds., *Media and Global Climate Knowledge: Journalism and the IPCC* (New York: Palgrave MacMillan, 2017).
14. Craig Calhoun, "Facets of the Public Sphere: Dewey, Arendt, Habermas," in *Institutional Change in the Public Sphere: Views on the Nordic Model*, ed. Fredrik Engelstad et al. (Berlin: De Gruyter, 2017), 23–45.
15. John Dewey, *The Public and Its Problems* (1927), ed. Melvin L. Rogers (Athens, GA: Swallow Press, 2016).

16. Jürgen Habermas, *The Structural Transformation of the Public Sphere: An Inquiry Into a Category of Bourgeois Society*, trans. Thomas Burger (Cambridge, MA: MIT Press, 1991).
17. Nancy Fraser, "Rethinking the Public Sphere: A Contribution to the Critique of Actually Existing Democracy," *Social Text*, nos. 25–26 (1990): 56–80.
18. Catherine R. Squires, "Rethinking the Black Public Sphere: An Alternative Vocabulary for Multiple Public Spheres," *Communication Theory* 12, no. 4 (2016): 446–68, https://doi.org/10.1111/j.1468-2885.2002.tb00278.x.
19. Hannah Arendt, *The Human Condition* (1958; reprint, Chicago: University of Chicago Press, 1998), 41. See also Habermas, *The Structural Transformation of the Public Sphere*.
20. Regina Marchi and Lynn Schofield Clark, "Social Media and Connective Journalism: The Formation of Counterpublics and Youth Civic Participation," *Journalism* 22, no. 2 (2021): 285–302.
21. Benedict Anderson, *Imagined Communities: Reflections on the Origin and Spread of Nationalism* (London: Verso, 1983).
22. Victor Pickard, "The Big Picture: Misinformation Society," *Public Books*, November 28, 2017, https://www.publicbooks.org/the-big-picture-misinformation-society/.
23. Scholars and commentators refer to climate denialism, skepticism, and contrarianism as forms of climate obstructionism. Skepticism and contrarianism are forms of denialism of the certainty of the causes and effects of the climate crisis; I therefore use the term *denialism* throughout the book.
24. Mark Hertsgaard and Kyle Pope, "The Media Are Complacent While the World Burns," *Columbia Journalism Review*, April 22 , 2019, https://www.cjr.org/special_report/climate-change-media.php.
25. Andrew Chadwick, *The Hybrid Media System: Politics and Power*, 2nd ed. (New York: Oxford University Press, 2017), 6.
26. See Manuel Castells, "An Introduction to the Information Age," *Colonial Latin American Review* 2, no. 7 (2007): 6–16, https://doi.org/10.1080/13604819708900050; Harry Cleaver, "The Zapatistas and the Electronic Fabric of Struggle," in *Zapatista! Reinventing Revolution in Mexico*, ed. John Holloway and Eloina Pelaez (Chicago: Pluto, 1998), 621–40.

27. Castells, "An Introduction."
28. Castells, "An Introduction."
29. Chadwick, *The Hybrid Media System*.
30. See, for example, Kate Crawford, *The Atlas of AI: Power, Politics, and the Planetary Costs of Artificial Intelligence* (New Haven, CT: Yale University Press, 2021), for a detailed view of the myth of clean tech and AI's contribution to the climate crisis.
31. Christopher M. Matthews, "Silicon Valley to Big Oil: We Can Manage Your Data Better Than You," *Wall Street Journal*, July 24, 2018, https://www.wsj.com/articles/silicon-valley-courts-a-wary-oil-patch-1532424600; Brian Merchant, "How Google, Microsoft, and Big Tech Are Automating the Climate Crisis," *Gizmodo*, February 21, 2019, https://gizmodo.com/how-google-microsoft-and-big-tech-are-automating-the-1832790799.
32. See "Google Cloud for Energy," Google, n.d., https://cloud.google.com/solutions/energy/.
33. See Tech Workers Coalition, "Climate Strike," n.d., https://techworkerscoalition.org/climate-strike/.
34. See, for example, the recent pledge by Meta CEO Mark Zuckerberg and his wife, Priscilla Chan, of $44 million in funding to support "solutions to climate change," most of which will go toward carbon-capture technology. Justine Calma, "Chan Zuckerberg Initiative Announces Tens of Millions in Funding for Climate Tech," *The Verge*, February 10, 2022, https://www.theverge.com/2022/2/10/22927245/chan-zuckerberg-initiative-millions-funding-climate-tech-carbon-removal.
35. InfluenceMap, "Big Tech and Climate Policy," January 2021, https://influencemap.org/report/Big-Tech-and-Climate-Policy-afb476c56f217ea0ab351d79096df04a.
36. Bernard C. Cohen, *The Press and Foreign Policy* (Princeton, NJ: Princeton University Press, 1963), 13.
37. Leah Lievrouw, "Materiality and Media in Communication and Technology Studies: An Unfinished Project," in *Media Technologies: Essays on Communication, Materiality, and Society*, ed. Tarleton Gillespie, Pablo J. Boczkowski, and Kirsten A. Foot (Cambridge, MA: MIT Press, 2014), 21–51.

38. Theodore Glasser, "Objectivity Precludes Responsibility," *The Quill* 72, no. 2 (1984): 13–16; Kunelius and Eide, "Moment of Hope, Mode of Realism."
39. See Michael Schudson, *Discovering the News: A Social History of American Newspapers* (New York: Basic, 1978); and Robert Karl Manoff and Michael Schudson, eds., *Reading the News: A Pantheon Guide to Popular Culture* (New York: Pantheon, 1986).
40. Glasser, "Objectivity Precludes Responsibility."
41. Maxwell T. Boykoff and Jules M. Boykoff, "Balance as Bias: Global Warming and the U.S. Prestige Press," *Global Environmental Change* 14, no. 2 (2004): 125–36. See also Maxwell T. Boykoff, *Who Speaks for the Climate? Making Sense of Media Reporting on Climate Change* (Cambridge: Cambridge University Press, 2011).
42. Michael Brüggemann, "Post-normal Journalism: Climate Journalism and Its Changing Contribution to an Unsustainable Debate," in *What Is Sustainable Journalism? Integrating the Environmental, Social, and Economic Challenges of Journalism*, ed. Peter Berglez, Ulrika Olausson, and Mart Ots (New York: Peter Lang, 2017), 57–73.
43. Edson C. Tandoc and Bruno Takahashi, "Playing a Crusader Role or Just Playing by the Rules? Role Conceptions and Role Inconsistencies Among Environmental Journalists," *Journalism* 15, no. 7 (2014): 889–907, https://doi.org/10.1177/1464884913501836; Adrienne Russell, "Innovation in Hybrid Spaces: 2011 UN Climate Summit and the Changing Journalism Field," *Journalism* 14, no. 7 (2013): 904–20.
44. James Painter et al., *Something Old, Something New: Digital Media and the Coverage of Climate Change* (Oxford: Reuters Institute for the Study of Journalism, 2016).
45. Simge Andı, "How People Access News About Climate Change," *Digital News Report*, March 2020, https://www.digitalnewsreport.org/survey/2020/how-people-access-news-about-climate-change.
46. Mike S. Schäfer and Inga Schlichting, "Media Representations of Climate Change: A Meta-analysis of the Research Field," *Environmental Communication* 8, no. 2 (2014): 142–60, https://doi.org/10.1080/17524032.2014.914050.
47. Julia Lück, Antal Wozniak, and Hartmut Wessler, "Networks of Coproduction: How Journalists and Environmental NGOs Create

Common Interpretations of the UN Climate Change Conferences," *International Journal of Press/Politics* 2, no. 1 (2016): 25–47, https://doi.org/10.1177/1940161215612204.

48. Jamie Henn, "Our Media Partnership with the *Guardian*," 350.org, April 1, 2015, https://350.org/our-media-partnership-with-the-guardian.
49. Rasmus Kleis Nielsen and Meera Selva, *More Important, but Less Robust? Five Things Everybody Needs to Know About the Future of Journalism*, Reuters Institute Report (Oxford: Reuters Institute for the Study of Journalism, January 2019), https://reutersinstitute.politics.ox.ac.uk/our-research/more-important-less-robust-five-things-everybody-needs-know-about-future-journalism.
50. IPCC, "About the IPCC," n.d., https://www.ipcc.ch/about/.
51. Chelsea Harvey, "New IPCC Report Looks at Neglected Element of Climate Action: People," "E&E News," *Scientific American*, April 7, 2022, https://www.scientificamerican.com/article/new-ipcc-report-looks-at-neglected-element-of-climate-action-people/.
52. Jim Skea quoted in Damian Carrington, "It's Over for Fossil Fuels: IPCC Spells Out What's Needed to Avert Climate Disaster," *The Guardian*, April 4, 2022, https://www.theguardian.com/environment/2022/apr/04/its-over-for-fossil-fuels-ipcc-spells-out-whats-needed-to-avert-climate-disaster.
53. Naomi Oreskes and Erik M. Conway, *Merchants of Doubt: How a Handful of Scientists Obscured the Truth on Issues from Tobacco Smoke to Global Warming* (New York: Bloomsbury, 2011), 17.
54. Oreskes and Conway, *Merchants of Doubt*, 17.
55. Candis Callison, *How Climate Change Comes to Matter: The Communal Life of Facts* (Durham, NC: Duke University Press, 2014), 3.
56. Callison, *How Climate Change Comes to Matter*, 2.
57. Callison, *How Climate Change Comes to Matter*, 9.
58. Heather Akin and Dietram A. Scheufele, "Overview of the Science of Science Communication," in *The Oxford Handbook of the Science of Science Communication*, ed. Katherine Hall Jamieson, Dan Kahan, and Dietram A. Scheufele (New York: Oxford University Press, 2017), 25–33.
59. Walter Lippmann, *Public Opinion* (New York: Harcourt, Brace, 1922).

60. Fred Turner, "The World Outside and the Pictures in Our Networks," in *Media Technologies*, ed. Gillespie, Boczkowski, and Foot, 251–60.
61. Deborah Lynn Guber, Jeremiah Bohr, and Riley E. Dunlap, "'TIME TO WAKE UP': Climate Change Advocacy in a Polarized Congress, 1996–2015," *Environmental Politics* 30, no. 4 (2020): 538–58, https://doi/10.1080/09644016.2020.1786333.
62. Guber, Bohr, and Dunlap quoted in Kate Yoder, "The Surprising Reasons Why People Ignore the Facts About Climate Change," *Grist*, July 28, 2020, https://grist.org/climate/the-surprising-reasons-why-people-ignore-the-facts-about-climate-change.
63. Erving Goffman, *Frame Analysis: An Essay on the Organization of Experience* (Cambridge, MA: Harvard University Press, 1974).
64. Robert M. Entman, "Framing: Toward Clarification of a Fractured Paradigm," *Journal of Communication* 43, no. 4 (1993): 52.
65. Matthew C. Nisbet, "Communicating Climate Change: Why Frames Matter for Public Engagement," *Environment: Science and Policy for Sustainable Development* 41, no. 2 (2009): 12– 23, https://doi.org/10.3200/ENVT.51.2.12-23.
66. Lorraine Whitmarsh and Adam Corner, "Tools for a New Climate Conversation: A Mixed-Methods Study of Language for Public Engagement Across the Political Spectrum," *Global Environmental Change* 42 (2017): 122–35, https://doi.org/10.1016/j.gloenvcha.2016.12.008; Teresa Myers et al., "A Public Health Frame Arouses Hopeful Emotions About Climate Change," *Climatic Change* 113, nos. 3–4 (2012): 1105–12, https://doi.org.10.1007/s10584-012-0513-6.
67. Mike Hulme, "Mediated Messages About Climate Change: Reporting the IPCC Fourth Assessment in the UK Print Media," *Climate Change and the Media* 2009:117–28.
68. David Wallace-Wells, "The Uninhabitable Earth," *New York Magazine*, July 9, 2017, https://nymag.com/intelligencer/2017/07/climate-change-earth-too-hot-for-humans.html.
69. Peter Neff quoted in Emmanuel Vincent, "Scientists Explain What *New York Magazine* Article on 'The Uninhabitable Earth' Gets Wrong," *Climate Feedback*, July 12, 2017, https://climatefeedback.org/evaluation/scientists-explain-what-new-york-magazine-article-on-the-uninhabitable-earth-gets-wrong-david-wallace-wells.

70. David Wallace-Wells quoted in Jennifer Szalai, "In 'The Uninhabitable Earth,' Apocalypse Is Now," *New York Times*, March 6, 2019, https://www.nytimes.com/2019/03/06/books/review-uninhabitable-earth-life-after-warming-david-wallace-wells.html.
71. Kate Aronoff, "Things Are Bleak! Jonathan Safran Foer's Quest for Planetary Salvation," *The Nation*, October 29, 2019, https://www.thenation.com/article/archive/jonathan-safran-foer-we-are-the-weather-climate-review.
72. Jonathan Watts, "Climatologist Michael E. Mann: 'Good People Fall Victim to Doomism. I Do Too Sometimes,'" *The Guardian*, February 27, 2021, https://www.theguardian.com/environment/2021/feb/27/climatologist-michael-e-mann-doomism-climate-crisis-interview.
73. Phaedra C. Pezzullo and Robert Cox, *Environmental Communication and the Public Sphere*, 5th ed. (Thousand Oaks, CA: SAGE, 2018), 258; Painter et al., *Something Old, Something New*.
74. Painter et al., *Something Old, Something New*.
75. Milane Larsson quoted in Painter et al., *Something Old, Something New*, 82.
76. Todd P. Newman, Erik C. Nisbet, and Matthew C. Nisbet, "Climate Change, Cultural Cognition, and Media Effects: Worldviews Drive News Selectivity, Biased Processing, and Polarized Attitudes," *Public Understanding of Science* 27, no. 8 (2018): 985–1002, https://doi.org/10.1177/0963662518801170.
77. Callison, *How Climate Change Comes to Matter*.
78. John Schwartz, "Katharine Hayhoe, a Climate Explainer Who Stays Above the Storm," *New York Times*, October 10, 2016, https://www.nytimes.com/2016/10/11/science/katharine-hayhoe-climate-change-science.html.
79. Maxwell T. Boykoff, *Creative (Climate) Communications: Productive Pathways for Science, Policy and Society* (Cambridge: Cambridge University Press, 2019), 211.
80. Chenjerai Kumanyika, tweet, Twitter, July 3, 2020, emphasis in original, https://twitter.com/catchatweetdown/status/1279110668214468612.
81. Whitney Phillips and Ryan M. Milner, *You Are Here: A Field Guide for Navigating Polarized Speech, Conspiracy Theories, and Our Polluted Media Landscape* (Cambridge, MA: MIT Press, 2021), 7.

82. Phillips and Milner, *You Are Here*, 4.
83. Boykoff, *Creative (Climate) Communications*, 20; Anthony Nadler, "Nature's Economy and News Ecology," *Journalism Studies* 20, no. 6 (2019): 823–39, https://doi.org/10.1080/1461670X.2018.1427000.
84. Whitney Phillips, "Navigating the Information Landscape: A Media Literacy Toolkit Series," *Commonplace*, June 30, 2020, https://doi.org/10.21428/6ffd8432.37155cc8.
85. Geoffrey C. Bowker and Susan Leigh Star, *Sorting Things Out: Classification and Its Consequences* (Cambridge, MA: MIT Press, 2000), 47.
86. Mike Ananny, *Networked Press Freedom: Creating Infrastructures for a Public Right to Hear* (Cambridge, MA: MIT Press, 2018), 111.
87. Frank A. Pasquale, "The Automated Public Sphere," University of Maryland Legal Studies Research Paper, no. 2017-31, https://ssrn.com/abstract=3067552.
88. Lam Thuy Vo, "Breaking Free from the Tyranny of the Loudest," NiemanLab blog, December 9, 2017, https://www.niemanlab.org/2017/12/breaking-free-from-the-tyranny-of-the-loudest.
89. Mike Ananny quoted in Emily J. Bell et al., "The Platform Press: How Silicon Valley Reengineered Journalism," Tow Center for Digital Journalism, March 29, 2017, https://www.cjr.org/tow_center_reports/platform-press-how-silicon-valley-reengineered-journalism.php.
90. Safiya Noble, *Algorithms of Oppression: How Search Engines Reinforce Racism* (New York: New York University Press, 2018), 3.
91. Joachim Allgaier, "Science and Environmental Communication on YouTube: Strategically Distorted Communications in Online Videos on Climate Change and Climate Engineering," *Frontiers in Communication* 4 (2019): 1–15, https://doi.org/10.3389/fcomm.2019.0036. See also María Carmen Erviti, José Azevedo, and Mónica Codina, "When Science Becomes Controversial," in *Communicating Science and Technology Through Online Video*, ed. Bienvenido León and Michael Bourk (London: Routledge, 2018), 46.
92. Jack Dorsey quoted in Lauren Jackson, "Jack Dorsey on Twitter's Mistakes," *New York Times*, August 7, 2020, https://www.nytimes.com/2020/08/07/podcasts/the-daily/Jack-dorsey-twitter-trump.html; for the podcast, see Michael Barbaro, "Jack Dorsey on Twitter's Mistakes,"

The Daily (podcast), *New York Times*, August 7, 2020, https://www.nytimes.com/2020/08/07/podcasts/the-daily/Jack-dorsey-twitter-trump.html.
93. Barbaro, "Jack Dorsey on Twitter's Mistakes."
94. For a rich and detailed account of how the idea of the computer was transformed from a threat during the Cold War into a means of achieving personal freedom, see Fred Turner, *From Counterculture to Cyberculture: Stewart Brand, the Whole Earth Network, and the Rise of Digital Utopianism* (Chicago: University of Chicago Press, 2006).
95. John Perry Barlow, "A Declaration of the Independence of Cyberspace" (1996), *Duke Law & Technology Review* 18, no. 1 (2019): 5–7.
96. Phillips and Milner, *You Are Here*, 50.
97. Phillips and Milner, *You Are Here*, 415.
98. Barbaro, "Jack Dorsey on Twitter's Mistakes."
99. Ananny, *Networked Press Freedom*, 2.

1. HOUSE ON FIRE

1. Isaac Stanley-Becker, "Trump, Pressed on the Environment in U.K. Visit, Says Climate Change Goes 'Both Ways,'" *Washington Post*, June 5, 2019, https://www.washingtonpost.com/world/europe/trump-pressed-on-the-environment-in-uk-visit-says-climate-change-goes-both-ways/2019/06/05/77c8750c-8717-11e9-9d73-e2ba6bbf1b9b_story.html.
2. Truth Tobacco Industry Documents, "Smoking and Health Proposal," n.d., https://www.industrydocuments.ucsf.edu/tobacco/docs/#id=psdw0147.
3. Justin Farrell, Kathryn McConnell, and Robert Brulle, "Evidence-Based Strategies for Combat Scientific Misinformation," *Nature Climate Change* 9 (2019): 191–95, https://doi.org/10.1038/s41558-018-0368-6.
4. Yanmengqian Zhou and Lijiang Shen, "Confirmation Bias and the Persistence of Misinformation on Climate Change," *Communication Research* 49, no. 4 (2022): 500–523.
5. Michael Schudson, *The Sociology of News* (New York: Norton, 2003).

6. Whitney Phillips, *The Oxygen of Amplification: Better Practices for Reporting on Extremists, Antagonists, and Manipulators Online* (New York: Data & Society Research Institute, 2018), https://datasociety.net/wp-content/uploads/2018/05/0-EXEC-SUMMARY_Oxygen_of_Amplification_DS-1.pdf.
7. James F. Black, "James Black Talk (1977)," *Inside Climate News*, September 15, 2015, https://insideclimatenews.org/documents/james-black-1977-presentation.
8. InfluenceMap, "Big Oil's Real Agenda on Climate Change," n.d., https://influencemap.org/report/How-Big-Oil-Continues-to-Oppose-the-Paris-Agreement-38212275958aa21196dae3b76220bddc.
9. Naomi Oreskes and Erik M. Conway, *Merchants of Doubt: How a Handful of Scientists Obscured the Truth on Issues from Tobacco Smoke to Climate Change* (New York: Bloomsbury, 2011).
10. David S. Ardia et al., "Addressing the Decline of Local News, Rise of Platforms, and Spread of Mis- and Disinformation Online: A Summary of Current Research and Policy Proposals," University of North Carolina Legal Studies Research Paper, 2020, https://citap.unc.edu/local-news-platforms-mis-disinformation/.
11. Ardia et al., "Addressing the Decline."
12. Sewall Chan, "Since 2005, Texas Has Lost More Newspaper Journalists per Capita Than All but Two Other States," *Texas Tribune*, June 29, 2022, https://www.texastribune.org/2022/06/29/death-local-news-texas/.
13. Penny Abernathy, "The State of Local News: The 2022 Report," Local News Initiative, June 29, 2022, https://localnewsinitiative.northwestern.edu/research/state-of-local-news/report/.
14. Molly Taft, "Chevron Jumps Into Texas' News Desert with Stories About Puppies, Football, and Oil," *Gizmodo*, August 18, 2022, https://gizmodo.com/chevron-local-news-texas-permian-proud-1849424317; Molly Taft. "For Earth Day, Houston Public Media Is Promoting . . . Chevron?," *Gizmodo*, April 21, 2022, https://gizmodo.com/houston-public-media-chevron-partnership-earth-day-1848822428.
15. Stephanie Craft and Morten S. Kristensen, "Noise and the Values of News," in *Rethinking Media Research for Changing Societies*, ed.

Matthew Powers and Adrienne Russell (Cambridge: Cambridge University Press, 2020), 78.

16. See Oreskes and Conway, *Merchants of Doubt*; W. Lance Bennett and Steven Livingstone, eds., *The Disinformation Age: Politics, Technology, and Disruptive Communication in the United States* (Cambridge: Cambridge University Press, 2021); Alice Marwick and Rebecca Lewis, "Media Manipulation and Disinformation Online," *Data & Society*, May 15, 2017, https://datasociety.net/library/media-manipulation-anddisinfo-online; Yochai Benkler, Robert Faris, and Hal Roberts, *Network Propaganda: Manipulation, Disinformation, and Radicalization in American Politics* (Oxford: Oxford Scholarship Online, 2018); and Lance Bennett and Barbara Pfetsch, "Rethinking Political Communication in a Time of Disrupted Public Spheres," *Journal of Communication* 68, no. 2 (2018): 243–53, https://doi.org/10.1093/joc/jqx017.

17. For updated tracking of the harassment of journalists, see the website of the Committee to Protect Journalists, https://cpj.org/.

18. Megan Brenan, "Media Confidence Ratings at Record Lows," Gallup, July 18, 2022, https://news.gallup.com/poll/394817/media-confidence-ratings-record-lows.aspx; and *Digital News Report 2022* (Oxford: Reuters Institute for the Study of Journalism, June 15, 2022), https://reutersinstitute.politics.ox.ac.uk/digital news-report/2022.

19. Ernesto Araújo quoted in Herton Escobar, "Brazil's New President Has Scientists Worried. Here's Why," *Science*, January 22, 2019, https://www.science.org/content/article/brazil-s-new-president-has-scientists-worried-here-s-why.

20. On Thierry Baudet, see Andrew Leigh, "How Populism Imperils the Planet," *MIT Press Reader*, November 5, 2021, https://thereader.mitpress.mit.edu/how-populism-imperils-the-planet/; for Trump's comment, see Donald Trump, tweet, Twitter, November 6, 2012, 11:15, https://twitter.com/realdonaldtrump/status/265895292191248385?lang=en.

21. Bill McKibbon, *CJR's Covering Climate Change*. YouTube video, May 17, 2019, 5:18:43, https://www.youtube.com/watch?v=FO9DKk07SCY.

22. John Hartley, *Understanding the News* (London: Methuen, 1982).

23. Maxwell Boykoff and Jules Boykoff, "Balance as Bias: Global Warming and the U.S. Prestige Press," *Global Environmental Change* 14, no. 2 (2004): 134.
24. Naomi Oreskes quoted in Amy Westervelt, "How the Fossil Fuel Industry Got the Media to Think Climate Change Was Debatable," *Washington Post*, January 10, 2019, https://www.washingtonpost.com/outlook/2019/01/10/how-fossil-fuel-industry-got-media-think-climate-change-was-debatable/.
25. Melissa Aronczyk and Maria I. Espinoza, *A Strategic Nature: Public Relations and the Politics of American Environmentalism* (Oxford: Oxford University Press, 2021), 8.
26. David Dickson, "The Case for a 'Deficit Model' of Science Communication," *Science and Development Network*, June 24, 2005, https://www.scidev.net/global/editorials/the-case-for-a-deficit-model-of-science-communic/.
27. Maxwell Boykoff and Jules Boykoff, "Climate Change and Journalistic Norms: A Case-Study of U.S. Mass-Media Coverage," *Geoforum* 38, no. 6 (2007): 1190–204, https://doi.org/10.1016/j.geoforum.2007.01.008.
28. danah boyd, "Media Manipulation, Strategic Amplification, and Responsible Journalism," *Data & Society: Points*, September 14, 2018, https://points.datasociety.net/media-manipulation-strategic-amplification-and-responsible-journalism-95f4d611f462.
29. Phillips, *The Oxygen of Amplification*, 2.
30. Mark Hertsgaard quoted in Kyle Pope, "Giving Climate the Coverage It Deserves," *Columbia Journalism Review*, June 15, 2022, https://www.cjr.org/covering_climate_now/giving-climate-the-coverage-it-deserves.php.
31. Extinction Rebellion, "XR NYC STANDARDS FOR MEDIA," n.d. [c. June 2019], https://www.xrebellion.nyc/media-standards.
32. For example, an article in the journal *BioScience* in January 2020, endorsed by more than eleven thousand scientists worldwide, declared "the climate crisis has arrived" and that efforts to conserve our biosphere need to vastly increase to avoid "untold suffering due to the climate crisis" (William Ripple et al., "World Scientists' Warning of a Climate Emergency," *BioScience* 70, no. 1 [January 2020]: 8–12).

33. James Painter et al., *Something Old, Something New: Digital Media and the Coverage of Climate Change* (Oxford: Reuters Institute for the Study of Journalism, 2016).
34. Nic Newman, with Richard Fletcher et al., *Digital News Report 2020* (Oxford: Reuters Institute for the Study of Journalism, 2020), https://reutersinstitute.politics.ox.ac.uk/sites/default/files/2020-06/DNR_2020_FINAL.pdf.
35. "The Sobering Realization That We're Going Completely in the Wrong Direction," *Media and Climate Change Observatory* 58 (October 2021), https://sciencepolicy.colorado.edu/icecaps/research/media_coverage/summaries/issue58.html.
36. Lucy McAllister et al., "Balance as Bias, Resolute on the Retreat? Updates and Analyses of Newspaper Coverage in the United States, United Kingdom, New Zealand, Australia and Canada Over the Past 15 Years," *Environmental Research Letters* 16, no. 9 (2021): 1–14, https://doi.org/10.1088/1748-9326/ac14eb.
37. Morgan McFall-Johnsen, "The Companies Polluting the Planet Have Spent Millions to Make You Think Carpooling and Recycling Will Save Us," *Business Insider*, September 18, 2021, https://www.businessinsider.com/fossil-fuel-companies-spend-millions-to-promote-individual-responsibility-2021-3; Mark Kaufman, "The Carbon Footprint Scam," *Mashable*, August 2021, https://mashable.com/feature/carbon-footprint-pr-campaign-sham.
38. Adrienne Russell et al., "Niche Climate News Sites and the Changing Context of Covering Catastrophe," *Journalism: Theory, Practice, Criticism*, online, July 6, 2022, https://doi.org/10.1177/14648849221113119.
39. Michael Brüggemann and Sven Engesser, "Between Consensus and Denial: Climate Journalists as Interpretive Community," *Science Communication* 36, no. 4 (2014): 399–427, https://doi.org/10.1177/1075547014533662.
40. Michael Brüggemann, "Post-normal Journalism: Climate Journalism and Its Changing Contribution to an Unsustainable Debate," in *What Is Sustainable Journalism? Integrating the Environmental, Social, and Economic Challenges of Journalism*, ed. Peter Berglez, Ulrika Olausson, and Mart Ots (New York: Peter Lang, 2017), 57–73.

41. "Exclusive: BBC Issues Internal Guidance on How to Report Climate Change," *Carbon Brief*, n.d., https://www.carbonbrief.org/exclusive-bbc-issues-internal-guidance-on-how-to-report-climate-change.
42. Robert Eshelman, discussion with the author, March 2016.
43. Matthew C. Nisbet, "Nature's Prophet: Bill McKibben as Journalist, Public Intellectual and Activist," Shornstein Center on Media, Politics, and Public Policy, March 7, 2013, shorensteincenter.org/wp-content/uploads/2013/03/D-78-Nisbet1.pdf.
44. Nesbit, "Nature's Prophet," 14.
45. Joy Jenkins and Lucas Graves, *Case Studies in Collaborative Local Journalism*, Reuters Institute Report (Oxford: Reuters Institute for the Study of Journalism, April 25, 2019), https://reutersinstitute.politics.ox.ac.uk/our-research/case-studies-collaborative-local-journalism.
46. Theodore L. Glasser, "Public Journalism Movement," in *The International Encyclopedia of Political Communication*, vol. 3, ed. Gianpietro Mazzoleni (Chichester, U.K.: Wiley Blackwell, 2015), https://onlinelibrary.wiley.com/doi/10.1002/9781118541555.wbiepc203.
47. Evlondo Cooper and Allison Fisher, "Covering Climate Now's Innovative Model Sets a New Standard for More and Better Climate Coverage," Media Matters, November 5, 2019, https://www.mediamatters.org/broadcast-networks/covering-climate-nows-innovative-model-sets-new-standard-more-and-better-climate.
48. Chris Outcalt and Brittany Peterson, "Colorado River 100 Years," *AP News*, September 12, 2022, https://apnews.com/hub/colorado-river-100-years.
49. Editorial quoted in Tom Zeller Jr., "Climate Talks Open with Calls for Urgent Action," *New York Times*, December 7, 2009, https://www.nytimes.com/2009/12/08/science/earth/08climate.html.
50. Leah Ceccarelli, "The Defense of Science in the Public Sphere," paper presented at the International Society for the Study of Argumentation, Amsterdam, July 3–6, 2018.
51. Norah MacKendrick, "Out of the Labs and Into the Streets: Scientists Get Political," *Sociological Forum* 32, no. 4 (2017): 896–902, https://doi.org/10.1111/socf.12366.

52. Leo Hickman quoted in Russell et al., "Niche Climate News Sites," 11.
53. Mat Hope quoted in Russell et al., "Niche Climate News Sites," 10.
54. Todd Gitlin, *The Whole World Is Watching* (Berkeley: University of California Press, 1980).
55. Hickman quoted in Russell et al., "Niche Climate News Sites," 11.
56. Vernon Loeb, interviewed by the author via Zoom, June 29, 2020.
57. Damien Carrington, "Why the *Guardian* Is Changing the Language It Uses About the Environment," *The Guardian*, May 17, 2019, https://www.theguardian.com/environment/2019/may/17/why-the-guardian-is-changing-the-language-it-uses-about-the-environment.
58. See, for example, Juliet Pinto, Robert E. Gutsche, and Paola Prado, eds., *Climate Change, Media & Culture: Critical Issues in Global Environmental Communication* (Bradford, U.K.: Emerald, 2019).
59. Adrienne Russell and Matt Tegelberg, "Beyond the Boundaries of Science: Resistance to Misinformation by Citizen Scientists," *Journalism: Theory, Practice, Criticism* 21, no. 3 (2019): 327–44, https://doi.org/10.1177/1464884919862655.
60. Rachel Ramirez, "Spanning Beats, Environmental Justice Reporting Influences Every Story," *NiemanReports*, February 3, 2021, https://niemanreports.org/articles/spanning-beats-environmental-justice-reporting-influences-every-story/.
61. Laurie Goering quoted in Russell et al., "Niche Climate News Sites," 13.
62. Loeb quoted in Russell et al., "Niche Climate News Sites," 14.
63. Paul J. Hirschfield and Simon Danella, "Legitimating Police Violence: Newspaper Narratives of Deadly Force," *Theoretical Criminology* 14, no. 2 (2010): 155–82, https://doi.org/10.1177/1362480609351545.
64. Allissa V. Richardson, *Bearing Witness While Black: African Americans, Smartphones, & the New Protest #Journalism* (New York: Oxford University Press, 2020), 43; see also Pamela J. Shoemaker, "Media Treatment of Deviant Political Groups," *Journalism Quarterly* 61, no. 1 (1984): 66–82.
65. Adrienne Russell, *Journalism as Activism: Recoding Media Power* (Cambridge: Polity, 2016).

66. W. Lance Bennett and Steven Livingston, "A Brief History of the Disinformation Age: Information Wars and the Decline of Institutional Authority," in *The Disinformation Age*, ed. Bennett and Livingston, 3.
67. Bennett and Livingston, "A Brief History of the Disinformation Age."
68. McFall-Johnsen, "The Companies Polluting the Planet."
69. For a detailed account of the asymmetrical relationship between news publishers and platforms, see Rasmus Kleis Nielsen and Sarah Anne Ganter, *The Power of Platforms: Shaping Media and Society* (Oxford: Oxford University Press, 2022).
70. Frank Pasquale, "The Automated Public Sphere," University of Maryland Legal Studies Research Paper, no. 2017-31, November 8, 2017, https://ssrn.com/abstract=3067552.
71. Yochai Benkler, *The Wealth of Networks: How Social Production Transforms Markets and Freedom* (New Haven, CT: Yale University Press, 2006); also see Manuel Castells, *Communication Power* (New York: Oxford University Press, 2009).
72. Zeynep Tufekci, "Facebook's Surveillance Machine," *New York Times*, March 19, 2018, https://www.nytimes.com/2018/03/19/opinion/facebook-cambridge-analytica.html.

2. NOISE, INCIVILITY, AND AMBIVALENCE

1. Ashley Gold, "Big Tech Antitrust Hearing," *Washington Journal*, C-SPAN, July 29, 2020, https://www.c-span.org/video/?474239-101/washington-journal-ashley-gold-big-tech-antitrust-hearing.
2. Cecilia Kang interviewed on Michael Barbaro, "The Big Tech Hearing," *The Daily* (podcast), *New York Times*, July 30, 2020, https://www.nytimes.com/2020/07/30/podcasts/the-daily/congress-facebook-amazon-google-apple.html.
3. See a video of the tobacco executives' testimony at https://www.youtube.com/watch?v=e_ZDQKq2Fo8.
4. Jonathan Tepper and Denise Hearn, *The Myth of Capitalism: Monopolies and the Death of Competition* (Newark, NJ: Wiley, 2018).
5. William Truvill, "The News 50: Tech Giants Dwarf Murdoch," *Press Gazette* (United Kingdom), December 3, 2020, https://pressgazette.co.uk/biggest-media-companies-world/.

6. Jean-Christophe Plantin and Aswin Punathambekar, "Digital Media Infrastructures: Pipes, Platforms, and Politics," *Media, Culture & Society* 41, no. 2 (2019): 163–74.
7. Mike Ananny, "Tech Platforms Are Where Public Life Is Increasingly Constructed, and Their Motivations Are Far from Neutral," NiemanLab blog, October 10, 2019, https://www.niemanlab.org/2019/10/tech-platforms-are-where-public-life-is-increasingly-constructed-and-their-motivations-are-far-from-neutral/.
8. Jürgen Habermas, *The Structural Transformation of The Public Sphere: An Inquiry Into a Category of Bourgeois Society*, trans. Thomas Burger (Cambridge, MA: MIT Press, 1991), 50.
9. For examples of such criticisms, see Catherine R. Squires, "Rethinking the Black Public Sphere: An Alternative Vocabulary for Multiple Public Spheres," *Communication Theory* 12, no. 4 (2016): 446–68; and Nancy Fraser, "Rethinking the Public Sphere: A Contribution to the Critique of Actually Existing Democracy," *Social Text*, nos. 25–26 (1990): 56–80.
10. Karin Wahl-Jorgensen, "Questioning the Ideal of the Public Sphere: The Emotional Turn," *Social Media + Society* 5, no. 3 (2019), https://doi.org/10.1177/2056305119852175.
11. See Jürgen Habermas, "Political Communication in Media Society: Does Democracy Still Enjoy an Epistemic Dimension? The Impact of Normative Theory on Empirical Research," *Communication Theory* 16, no. 4 (2006): 411–26; and Hartmut Wessler, *Habermas and the Media* (London: Wiley, 2019).
12. Hannah Arendt, *The Human Condition* (1958; reprint, Chicago: University of Chicago Press, 1998), 198.
13. Evlondo Cooper, "How Broadcast TV News Covered Environmental Justice in 2021," Media Matters, May 10, 2022, https://www.mediamatters.org/broadcast-networks/how-broadcast-tv-news-covered-environmental-justice-2021.
14. Adrienne Russell et al., "Culture: Media Convergence and Networked Participation," in *Networked Publics*, ed. Kazys Varnelis (Cambridge, MA: MIT Press, 2008), 43.
15. Yochai Benkler, *The Wealth of Networks: How Social Production Transforms Markets and Freedom* (New Haven, CT: Yale University Press, 2006).

16. Matt Carlson, *Journalistic Authority: Legitimating News in the Digital Era* (New York: Columbia University Press, 2017).
17. Seth C. Lewis and Logan Molyneux, "A Decade of Research on Social Media and Journalism: Assumptions, Blind Spots, and a Way Forward," *Media and Communication* 6, no. 4 (2018): 12.
18. Adrienne Russell, *Journalism as Activism: Recoding Media Power* (Cambridge: Polity, 2016).
19. Zizi Papacharissi, *Affective Publics: Sentiment, Technology and Politics* (Oxford: Oxford University Press, 2015), 117.
20. Papacharissi, *Affective Publics*, 117.
21. Peter Dahlgren, "Media, Knowledge and Trust: The Deepening Epistemic Crisis of Democracy," *Javnost–The Public* 25, nos. 1–2 (2018): 20–27.
22. Daniel Kreiss, *Prototype Politics: Technology-Intensive Campaigning and the Data of Democracy* (Oxford: Oxford University Press, 2016).
23. Frank A. Pasquale, "The Automated Public Sphere," University of Maryland Legal Studies Research Paper, no. 2017-31, 2017, https://ssrn.com/abstract=3067552.
24. Bruce Bimber and Homero Gil de Zúñiga, "The Unedited Public Sphere," *New Media & Society* 22, no. 4 (2020): 700–715, https://doi.org/10.1177/1461444819893980.
25. Claire Wardle and Hossein Derakhshan, "Thinking About 'Information Disorder': Formats of Misinformation, Disinformation, and Mal-Information," in *Journalism, "Fake News" & Disinformation: Handbook for Journalism Education and Training*, ed. Cherilyn Ireton and Julie Posett (Paris: UNESCO, 2018), 43–54.
26. W. Lance Bennett and Barbara Pfetsch, "Rethinking Political Communication in a Time of Disrupted Public Spheres," *Journal of Communication* 68, no. 2 (2018): 243–53, https://doi.org/10.1093/joc/jqx017.
27. Noortje Marres, "Why We Can't Have Our Facts Back," *Engaging Science, Technology, and Society* 4 (2018): 423–43, https://doi.org/10.17351/ests2018.188.
28. Whitney Phillips and Ryan M. Milner, *You Are Here: A Field Guide for Navigating Polarized Speech, Conspiracy Theories, and Our Polluted Media Landscape* (Cambridge, MA: MIT Press, 2021).
29. Ethan Zuckerman, "What Is Digital Public Infrastructure?," Center for Journalism & Liberty, November 17, 2020, https://www.jou

nalismliberty.org/publications/what-is-digital-public-infrastructure#_ednref3.

30. Tarleton Gillespie, "Governance of and by Platforms," in *The SAGE Handbook of Social Media*, ed. Jean Burgess, Thomas Poell, and Alice Marwick (Los Angeles: SAGE, 2017), 254–78.
31. See Arne Hintz, Lina Dencik, and Karin Wahl-Jorgensen, *Digital Citizenship in a Datafied Society* (London: Wiley, 2018); Joseph Turow, *The Daily You: How the New Advertising Industry Is Defining Your Identity and Your Worth* (New Haven, CT: Yale University Press, 2012); and Shoshana Zuboff, *The Age of Surveillance Capitalism: The Fight for a Human Future at the New Frontier of Power* (London: Profile, 2019).
32. Arne Hintz, Lina Dencik, and Karin Wahl-Jorgensen, "Introduction," in "Digital Citizenship and Surveillance," special issue of *International Journal of Communication* 11 (2017): 731–39, https://doi.org/1932-8036/20170005.
33. Zuboff, *The Age of Surveillance Capitalism*.
34. Hintz, Dencik, and Wahl-Jorgensen, "Introduction," 732.
35. Victor Åström, "Environmental Report Defenders Under Attack," Swedish Society for Nature Conservation, November 2019, 22, https://www.universal-rights.org/wp-content/uploads/2019/11/environmental_defenders_under_attack_eng.pdf.
36. Jenna Bitar, "6 Ways Government Is Going After Environmental Activists," *ACLU News*, February 6, 2018, https://www.aclu.org/news/free-speech/6-ways-government-going-after-environmental-activists.
37. Alleen Brown, Will Parrish, and Alice Speri, "Leaked Documents Reveal Counterterrorism Tactics Used at Standing Rock to 'Defeat Pipeline Insurgencies,'" *The Intercept* (blog), May 27, 2017, https://theintercept.com/2017/05/27/leaked-documents-reveal-security-firms-counterterrorism-tactics-at-standing-rock-to-defeat-pipeline-insurgencies/.
38. Steve Handley, "Bill McKibben Calls FBI Tracking of Environmental Activists 'Contemptible,'" CleanTechnica, December 13, 2018, https://cleantechnica.com/2018/12/13/bill-mckibben-calls-fbi-tracking-of-environmental-activists-contemptible/.

39. For a detailed report on the use of the classification "terrorist" and how the fossil-fuel industry has supported the targeting and restricting of climate protesters in the United States, see Grace Nosek, "The Fossil Fuel Industry's Push to Target Climate Protesters in the U.S.," *Pace Environmental Law Review* 38 (2020), https://digitalcommons.pace.edu/pelr/vol38/iss1/2.
40. See Julia Angwin et al., "Machine Bias," *ProPublica*, May 23, 2016, https://www.propublica.org/article/machine-bias-risk-assessments-in-criminal-sentencing; and Daniel Trottier and Christian Fuchs, "Theorising Social Media, Politics and the State: An Introduction," in *Social Media, Politics and the State: Protests, Revolutions, Riots, Crime and Policing in the Age of Facebook, Twitter, and YouTube*, ed. Daniel Trottier and Christian Fuchs (New York: Routledge, 2015), 15–50.
41. Karen Li Xan Wong and Amy Shields Dobson, "We're Just Data: Exploring China's Social Credit System in Relation to Digital Platform Ratings Cultures in Westernised Democracies," *Global Media and China* 4, no. 2 (2019): 220–32, https;//doi.org/10.1177/2059436419856090.
42. Safiya Noble, interview, USC Annenberg, School for Communication and Journalism, "In 'Algorithms of Oppression,' Safiya Noble Finds Old Stereotypes Persist in New Media," February 16, 2018, updated December 11, 2019, https://annenberg.usc.edu/news/diversity-and-inclusion/algorithms-oppression-safiya-noble-finds-old-stereotypes-persist-new. See also Safiya Noble, *Algorithms of Oppression: How Search Engines Reinforce Racism* (New York: New York University Press, 2018).
43. Joseph Turow and Nick Couldry, "Media as Data Extraction: Towards a New Map of a Transformed Communications Field," *Journal of Communication* 68, no. 2 (2018): 415–23, https://doi.org/10.1093/joc/jqx011.
44. InfluenceMap, "Climate Change and Digital Advertising: Climate Science Disinformation in Facebook Advertising," October 2020, https://influencemap.org/report/Climate-Change-and-Digital-Advertising-86222daed29c6f49ab2da76b0df15f76#4. Groups identified by the InfluenceMap report include PragerU, the Mackinac Center for

Public Policy, the Texas Public Policy Foundation, and the Competitive Enterprise Institute, among others.

45. Carole Cadwalladr and Emma Graham-Harrison, "Revealed: 50 Million Facebook Profiles Harvested for Cambridge Analytica in Major Data Breach," *The Guardian*, March 17, 2018, https://www.theguardian.com/news/2018/mar/17/cambridge-analytica-facebook-influence-us-election.

46. Samantha Bradshaw and Philip Howard, "Troops, Trolls and Troublemakers: A Global Inventory of Organized Social Media Manipulation," Oxford Internet Institute, 2017, https://ora.ox.ac.uk/objects/uuid:cef7e8d9-27bf-4ea5-9fd6-855209b3e1f6.

47. Eli Pariser, *The Filter Bubble: How the New Personalized Web Is Changing What We Read and How We Think* (New York: Penguin, 2011).

48. Cass R. Sunstein, *#Republic: Divided Democracy in the Age of Social Media* (Princeton, NJ: Princeton University Press, 2018).

49. Chris Bail, *Breaking the Social Media Prism* (Princeton, NJ: Princeton University Press, 2021).

50. Birgit Stark et al., "Are Algorithms a Threat to Democracy? The Rise of Intermediaries: A Challenge for Public Discourse," Algorithm Watch, May 2020, https://algorithmwatch.org/en/wp-content/uploads/2020/05/Governing-Platforms-communications-study-Stark-May-2020-AlgorithmWatch.pdf.

51. Herbert Simon quoted in Matthew Hindman, *The Internet Trap: How the Digital Economy Builds Monopolies and Undermines Democracy* (Princeton, NJ: Princeton University Press, 2018), 4.

52. Maria Cantwell quoted in U.S. Senate, Committee on Commerce, Science, and Transportation, "At Hearing with Big Tech CEOs, Cantwell Defends Local Journalism, Presses Platforms on Unfair Practices," October 2020, https://www.commerce.senate.gov/2020/10/at-hearing-with-big-tech-ceos-cantwell-defends-local-journalism-presses-platforms-on-unfair-practices.

53. Nancy Dubuc quoted in Jessica Bursztynsky, "Vice Media CEO Slams Big Tech as 'Great Threat to Journalism' in Layoffs Memo," CNBC, May 15, 2020, https://www.cnbc.com/2020/05/15/vice-media-ceo-slams-big-tech-as-great-threat-to-journalism.html.

54. Victor Pickard, "Restructuring Democratic Infrastructures: A Policy Approach to the Journalism Crisis," *Digital Journalism* 8, no. 6 (2020): 704–19, https://doi.org/10.1080/21670811.2020.1733433.
55. Nick Couldry, "The Myth of 'Us': Digital Networks, Political Change and the Production of Collectivity," *Information, Communication & Society* 18, no. 6 (2015): 623.
56. Also see Nick Couldry and Ulises A. Mejias, *The Costs of Connection: How Data Are Colonizing Human Life and Appropriating It for Capitalism* (Palo Alto, CA: Stanford University Press, 2019).
57. Marres, "Why We Can't Have Our Facts Back," 435.
58. Chris Anderson, "The End of Theory: The Data Deluge Makes the Scientific Method Obsolete," *Wired*, June 23, 2008, https://www.wired.com/2008/06/pb-theory/.
59. Melissa Aronczyk, "Public Communication in a Promotional Culture," in *Rethinking Media Research for Changing Societies*, ed. Matthew Powers and Adrienne Russell (Cambridge: Cambridge University Press, 2020), 39.
60. Bail, *Breaking the Social Media Prism*, 10.
61. Joachim Allgaier, "Science and Environmental Communication on YouTube: Strategically Distorted Communications in Online Videos on Climate Change and Climate Engineering," *Frontiers in Communication* 4 (2019): 36, https://doi.org/10.3389/fcomm.2019.00036.
62. Dustin Welbourne and Will J. Grant, "What Makes a Popular Science Video on YouTube," *The Conversation*, February 24, 2015, http://theconversation.com/what-makes-a-popular-science-videoon-youtube-36657.
63. María Carmen Erviti, José Azevedo, and Mónica Codina, "When Science Becomes Controversial," in *Communicating Science and Technology Through Online Video*, ed. Bienvenido León and Michael Bourk (London: Routledge, 2018), 46.
64. Allgaier, "Science and Environmental Communication on YouTube," abstract.
65. Bernhard Rieder, Ariadna Matamoros-Fernández, and Òscar Coromina, "From Ranking Algorithms to 'Ranking Cultures': Investigating the Modulation of Visibility in YouTube Search Results," *Convergence* 24, no. 1 (2018): 64, https://doi.org/10.1177/1354856517736982.

66. Zeynep Tufekci, "YouTube, the Great Radicalizer," *New York Times*, March 10, 2018, https://www.nytimes.com/2018/03/10/opinion/sunday/youtube-politics-radical.html.
67. Pasquale, "The Automated Public Sphere," 3.
68. See Aaron M. McCright and Riley E. Dunlap, "The Politicization of Climate Change and Polarization in the American Public's Views of Global Warming, 2001–2010," *Sociological Quarterly* 52, no. 2 (2011): 155–94, https://doi.org/10.1111/j.1533-8525.2011.01198; and Robert J. Brulle, "Institutionalizing Delay: Foundation Funding and the Creation of U.S. Climate Change Counter-Movement Organizations," *Climatic Change* 122, no. 4 (2014): 681–94, https://doi.org/10.1007/s10584-013-1018-7.
69. Christopher Knaus, "Bots and Trolls Spread False Arson Claims in Australian Fires 'Disinformation Campaign,'" *The Guardian*, January 7, 2020, https://www.theguardian.com/australia-news/2020/jan/08/twitter-bots-trolls-australian-bushfires-social-media-disinformation-campaign-false-claims.
70. Jason Wilson, "Social Media Disinformation on U.S. West Coast Blazes 'Spreading Faster Than Fire,'" *The Guardian*, September 14, 2020, https://www.theguardian.com/us-news/2020/sep/14/disinformation-oregon-wildfires-spreading-social-media.
71. FBI Portland, "Statement on Misinformation Related to Wildfires," September 11, 2020, https://www.fbi.gov/contact-us/field-offices/portland/news/press-releases/fbi-releases-statement-on-misinformation-related-to-wildfires.
72. Christopher Bouzy quoted in Marianne Lavelle, "'Trollbots' Swarm Twitter with Attacks on Climate Science Ahead of UN Summit," *Inside Climate News*, September 16, 2019, https://insideclimatenews.org/news/16092019/trollbot-twitter-climate-change-attacks-disinformation-campaign-mann-mckenna-greta-targeted/.
73. Ryan Brooks, "How Russians Attempted to Use Instagram to Influence Native Americans," *Buzzfeed*, October 23, 2017, https://www.buzzfeednews.com/article/ryancbrooks/russian-troll-efforts-extended-to-standing-rock#.feJrJXRJ.
74. Sara C. Francisco and Diane H. Felmlee. "What Did You Call Me? An Analysis of Online Harassment Towards Black and Latinx Women," *Race and Social Problems* 14, no. 1 (2022): 1–13; Marjan Nadim

2. NOISE, INCIVILITY, AND AMBIVALENCE ⌘ 191

and Audun Fladmoe, "Silencing Women? Gender and Online Harassment," *Social Science Computer Review* 39, no. 2 (2021): 245–58, https://doi.org/10.1177/0894439319865518.

75. Jacob Nelson, "A Twitter Tightrope Without a Net: Journalists' Reactions to Newsroom Social Media Policies," *Columbia Journalism Review*, December 2, 2021, https://www.cjr.org/tow_center_reports/newsroom-social-media-policies.php.
76. Peter Schwartzstein, "The Authoritarian War on Environmental Journalism," Century Foundation, July 7, 2020, https://tcf.org/content/report/authoritarian-war-environmental-journalism/; Eric Freedman, "Environmental Journalists Face Threats, Violence," International Journalists Network, November 30, 2018, https://ijnet.org/en/story/environmental-journalists-face-threats-violence.
77. Naomi Oreskes, "Beyond the Ivory Tower: The Scientific Consensus on Climate Change," *Science* 306, no. 5702 (2004): 1686, https://doi.org/10.1126/science.1103618.
78. Candis Callison, *How Climate Change Comes to Matter: The Communal Life of Facts* (Durham, NC: Duke University Press, 2015), 29.
79. Silvio Waisbord, "Antipress Violence and the Crisis of the State," *Harvard International Journal of Press/Politics* 7, no. 3 (2002): 90–109, https://doi.org/10.1177/1081180X0200700306.
80. Seth C. Lewis, Rodrigo Zamith, and Mark Coddington, "Online Harassment and Its Implications for the Journalist–Audience Relationship," *Digital Journalism* 8, no. 8 (2020): 1047–67, https://doi.org/10.1080/21670811.2020.1811743.
81. Scott Waldman and Niina Heikkinen, "As Climate Scientists Speak Out, Sexist Attacks Are on the Rise," "E&E News," *Scientific American*, August 22, 2018, https://www.scientificamerican.com/article/as-climate-scientists-speak-out-sexist-attacks-are-on-the-rise/.
82. Katherine Hayhoe quoted in Katherine Bagley and Naveena Sadasivam, "Climate Denial's Ugly Side: Hate Mail to Scientists," *Inside Climate News*, December 11, 2015, https://insideclimatenews.org/news/11122015/climate-change-global-warming-denial-ugly-side-scientists-hate-mail-hayhoe-mann.
83. Arron Banks, tweet, Twitter, August 14, 2019, 2:11, https://twitter.com/arron_banks/status/1161747086616010752?lang=en.

84. "Catherine McKenna: Canada Environment Minister Given Extra Security," *BBC News*, September 8, 2019, https://www.bbc.com/news/world-us-canada-49627153.
85. Ashley A. Anderson et al., "The 'Nasty Effect': Online Incivility and Risk Perceptions of Emerging Technologies," *Journal of Computer-Mediated Communication* 19, no. 3 (2014): 383, http://doi.org/10.1111/jcc4.12009.
86. Dominique Brossard and Dietram A. Scheufele, "Science, New Media, and the Public," *AAAS*, January 4, 2013, http://www.sciencemag.org/content/339/6115/40.short.
87. Whitney Phillips and Ryan M. Milner, *The Ambivalent Internet: Mischief, Oddity, and Antagonism Online* (London: Wiley, 2018).
88. Benkler, *The Wealth of Networks*.
89. Henry Jenkins, *Convergence Culture: Where Old and New Media Collide* (Cambridge, MA: MIT Press, 2006).
90. Axel Bruns, *Blogs*, Wikipedia, Second Life, *and Beyond: From Production to Produsage*, Digital Formations, vol. 45 (New York: Peter Lang, 2008).
91. Varnelis, *Networked Publics*.
92. Ashley Hedrick, Dave Karpf, and Daniel Kreiss, "The Earnest Internet vs. the Ambivalent Internet," *International Journal of Communication* 12 (2018): 1057–64.
93. Phillips and Milner, *The Ambivalent Internet*, 25.
94. Craig Timberg and Tony Romm, "Facebook CEO Mark Zuckerberg to Capitol Hill: 'It Was My Mistake, and I'm Sorry,'" *Washington Post*, April 9, 2018, https://www.washingtonpost.com/news/the-switch/wp/2018/04/09/facebook-chief-executive-mark-zuckerberg-to-captiol-hill-it-was-my-mistake-and-im-sorry/.
95. The conversation is quoted in Jose Antonio Vargas, "The Face of Facebook," *The New Yorker*, September 13, 2010, https://www.newyorker.com/magazine/2010/09/20/the-face-of-facebook.
96. Hedrick, Karpf, and Kreiss, "The Earnest Internet vs. the Ambivalent Internet," 1058.
97. Climate Science Information Center, "How Much Do You Know About Climate Change?," Facebook, n.d. [c. September 2020], https://www.facebook.com/climatescienceinfo.

2. NOISE, INCIVILITY, AND AMBIVALENCE ⌘ 193

98. Edward Palmieri, "The Next Decade: How Facebook Is Stepping Up the Fight Against Climate Change," *Facebook Engineering*, September 14, 2020, https://engineering.fb.com/2020/09/14/data-center-engineering/net-zero-carbon.
99. Paul Griffin, *The Carbon Majors Database: CDP Carbon Majors Report 2017* (Snowmass, CO: CDP, Climate Accountability Institute, July 2017), 8 https://cdn.cdp.net/cdp-production/cms/reports/documents/000/002/327/original/Carbon-Majors-Report-2017.pdf?1501833772.
100. Afiq Fitri, "The Tech Industry's Progress on Carbon Emissions Has Been Mixed," TechMonitor, June 23, 2022, https://techmonitor.ai/leadership/sustainability/tech-industry-carbon-emissions-progress.
101. Brian Kahn, "Is This a Joke?," *Gizmodo*, September 15, 2020, https://earther.gizmodo.com/is-this-a-joke-1845063054.
102. Scott Waldman, "Climate Denial Spreads on Facebook as Scientists Face Restrictions," *ClimateWire*, July 6, 2020, https://www.scientificamerican.com/article/climate-denial-spreads-on-facebook-as-scientists-face-restrictions/.
103. Scott Waldman, "How CO_2 Boosters' Op-Ed Slipped by Facebook Fact-Checkers," *ClimateWire*, June 23, 2020, https://www.eenews.net/stories/1063436369.
104. Caleb Rossiter and CO_2 Coalition ad quoted in Waldman, "How CO_2 Boosters' Op-Ed."
105. Shoshana Zuboff, "The Coup We Are Not Talking About," *New York Times*, January 29, 2020, https://www.nytimes.com/2021/01/29/opinion/sunday/facebook-surveillance-society-technology.html.
106. Andrew Bosworth quoted in Zuboff, "The Coup We Are Not Talking About."
107. Anusha Narayanan quoted in Brett Wilkins, "Green Groups' Petition Urges Social Media Platforms to Ban Big Oil Ads," Common Dreams, June 28, 2021, https://www.commondreams.org/news/2021/06/28/green-groups-petition-urges-social-media-platforms-ban-big-oil-ads.
108. Harriet Kingaby, "Climate Disinformation: A Beginner's Guide," *Branch*, no. 2 (Spring 2021), https://branch.climateaction.tech/issues/issue-2/climate-disinformation-a-beginners-guide/.

109. Marres, "Why We Can't Have Our Facts Back."
110. Marres, "Why We Can't Have Our Facts Back," 435.
111. Marres, "Why We Can't Have Our Facts Back," 435.

3. AFTER PEAK INDIFFERENCE

1. Clive Thompson, "We Might Be Reaching 'Peak Indifference' on Climate Change," *Wired*, March 25, 2019, https://www.wired.com/story/we-might-be-reaching-peak-indifference-on-climate-change/.
2. Adam B. Smith, "2020 U.S. Billion-Dollar Weather and Climate Disasters in Historical Context," Climate.gov, 2021, https://www.climate.gov/disasters2020.
3. Anthony Leiserowitz et al., *Climate Change in the American Mind: December 2020* (New Haven, CT: Yale Program on Climate Change Communication, February 10, 2021), https://climatecommunication.yale.edu/publications/climate-change-in-the-american-mind-december-2020/.
4. James Bell et al., "In Response to Climate Change, Citizens in Advanced Economies Are Willing to Alter How They Live and Work," Pew Research Center, September 14, 2021, https://www.pewresearch.org/global/2021/09/14/in-response-to-climate-change-citizens-in-advanced-economies-are-willing-to-alter-how-they-live-and-work/.
5. Robert Gifford, "The Dragons of Inaction: Psychological Barriers That Limit Climate Change Mitigation and Adaptation," *American Psychologist* 66, no. 4 (2011): 290–302, https://doi.org/10.1037/a0023566.
6. Matthew Ballew et al., "Which Racial/Ethnic Groups Care Most About Climate Change?," Yale Program on Climate Change Communication, April 16, 2020, https://climatecommunication.yale.edu/publications/race-and-climate-change/.
7. Christopher W. Tessum et al., "PM2.5 Polluters Disproportionately and Systemically Affect People of Color in the United States," *ScienceAdvances* 7, no. 18 (2021): 1–6, https://doi.org/10.1126/sciadv.abf4491; Hiroko Tabuchi and Nadja Popovich, "People of Color Breathe More Hazardous Air. The Sources Are Everywhere," *New*

York Times, April 28, 2021, https://www.nytimes.com/2021/04/28/climate/air-pollution-minorities.html.
8. Zahra Hirji, "It's Not Just Greta. Trolls Are Swarming Young Climate Activists Online," BuzzFeed News, September 25, 2019, https://www.buzzfeednews.com/article/zahrahirji/greta-thunberg-climate-teen-activist-harassment.
9. Cary Funk, "Key Findings: How Americans' Attitudes About Climate Change Differ by Generation, Party and Other Factors," Pew Research Center blog, May 26, 2021, https://www.pewresearch.org/fact-tank/2021/05/26/key-findings-how-americans-attitudes-about-climate-change-differ-by-generation-party-and-other-factors/. Members of GenZ were born after 1996; millennials between 1981 and 1996; GenXers between 1965 and 1980; Boomers and older before 1965.
10. Ann V. Sanson, Judith Van Hoorn, and Susie E. L. Burke, "Responding to the Impacts of the Climate Crisis on Children and Youth," *Child Development Perspectives* 13, no. 4 (2019): 201–7, https://doi.org/10.1111/cdep.12342.
11. Walter Lippmann, *The Phantom Public: A Sequel to "Public Opinion"* (New York: Macmillan, 1927), 11.
12. Some argue the opposite: that publics simply get in the way of climate solutions. For example, proponents of eco-authoritarianism argue for the inevitability and necessity of measures that defy democracy and even commonly accepted human rights in general. See Democracia Abierta, "Democracy, Authoritarianism and the Climate Emergency," *openDemocracy*, September 18, 2019, https://www.opendemocracy.net/en/democraciaabierta/democracia-autoritarismo-y-emergencia-clim%C3%A1tica-en/.
13. Noortje Marres, "Issues Spark a Public Into Being: A Key but Often Forgotten Point of the Lippmann–Dewey Debate," in *Making Things Public: Atmospheres of Democracy* (Cambridge, MA: MIT Press, 2005), 217.
14. For an overview of how the fossil-fuel economy is integral to the current world order, see chapters 9 and 10 of Amitav Ghosh, *The Nutmeg's Curse: Parables for a Planet in Crisis* (Chicago: University of Chicago Press, 2021).

15. Nancy Fraser, *Scales of Justice: Reimagining Political Space in a Globalizing World* (New York: Columbia University Press, 2018), 16.
16. Anna Roosvall and Matt Tegelberg, *Media and Transnational Climate Justice Indigenous Activism and Climate Politics* (New York: Peter Lang, 2017).
17. Guy Rez, "How I Built Resilience: Varshini Prakash of Sunrise Movement," *How I Built This* (podcast), NPR, November 5, 2020, https://www.npr.org/2020/10/28/928810660/how-i-built-resilience-varshini-prakash-of-sunrise-movement.
18. Christina Neumayer, Mette Mortensen, and Thomas Poell, "Introduction: Social Media Materialities and Protest," in *Social Media Materialities and Protest: Critical Reflections*, ed. Mette Mortensen, Christina Neumayer, and Thomas Poell (London: Routledge, 2019), 1–14.
19. Zeynep Tufekci, *Twitter and Tear Gas: The Power and Fragility of Networked Protest* (New Haven, CT: Yale University Press, 2017).
20. Veronica Barassi, *Activism on the Web: Everyday Struggles Against Digital Capitalism* (London: Routledge, 2015).
21. Lina Dencik, Arne Hintz, and Jonathan Cable, "Towards Data Justice? The Ambiguity of Anti-surveillance Resistance in Political Activism," *Big Data & Society* 3, no. 2 (December 2016): 1–12, https://doi.org/10.1177/2053951716679678.
22. Emiliano Treré, *Hybrid Media Activism: Ecologies, Imaginaries, Algorithms* (New York: Routledge, 2018).
23. Stefania Milan, "The Materiality of Clouds: Beyond a Platform-Specific Critique of Contemporary Activism," in *Social Media Materialities and Protest*, ed. Mortensen, Neumayer, and Poell, 116–27.
24. Matthew Taylor, "Environment Protest Being Criminalised Around World, Say Experts," *The Guardian*, April 29, 2021, https://www.theguardian.com/environment/2021/apr/19/environment-protest-being-criminalised-around-world-say-experts.
25. Miren Gutiérrez, *Data Activism and Social Change* (London: Palgrave Pivot, 2018).
26. Bart Cammaerts, "Protest Logics and the Mediation Opportunity Structure," *European Journal of Communication* 27, no. 2 (June 2012): 117–34, https://doi.org/10.1177/0267323112441007.

27. Greta Thunberg et al., "This Is the World Being Left to Us by Adults," *New York Times*, August 19, 2021, https://www.nytimes.com/2021/08/19/opinion/climate-un-report-greta-thunberg.html.
28. "Reasons to Strike," Fridays for Future blog, n.d., https://fridaysforfuture.org/take-action/reasons-to-strike/.
29. Maria Reyes and Adriana Calderón, "What Is MAPA and Why Should We Pay Attention to It?," Fridays for Future blog, March 13, 2021, https://fridaysforfuture.org/newsletter/edition-no-1-what-is-mapa-and-why-should-we-pay-attention-to-it/.
30. Bart Cammaerts, "The Mediated Circulation of the UK Youth-Strike4Climate Movement's Discourses and Actions," *European Journal of Cultural Studies* (forthcoming).
31. Whitney Jefferson, "Here's How Celebrities Are Taking Part in the Climate Strike," *BuzzFeed*, November 12, 2021, https://www.buzzfeed.com/whitneyjefferson/celebrities-taking-part-in-the-climate-strike; "Act Today, Save Tomorrow," Patagonia, n.d., https://eu.patagonia.com/gb/en/climatestrike/.
32. Felix Brünker, Fabian Deitelhoff, and Milad Mirbabaie, "Collective Identity Formation on Instagram—Investigating the Social Movement Fridays for Future," paper presented at the Australasian Conference on Information Systems, Perth, December 11, 2019, http://arxiv.org/abs/1912.05123.
33. Emma Pattee, "Meet the Climate Change Activists of TikTok," *Wired*, March 11, 2021, https://www.wired.com/story/climate-change-tiktok-science-communication/. See also the Instagram account @clime mechange.
34. Ellen Scott, "Meet the Teens Making Climate Change Memes to Deal with Ecoanxiety," *Metro* (blog), August 15, 2019, https://metro.co.uk/2019/08/15/meet-teens-making-climate-change-memes-deal-ecoanxiety-10570574/.
35. Global Climate Strike, "Guides and Trainings," n.d., https://globalclimatestrike.net/guides/.
36. Allie Rougeot, interviewed by Matt Tegelberg, Toronto, September 16, 2019, transcript shared with the author.
37. Grace Lambert, interviewed by the author, Seattle, November 3, 2019; all quotations from Lambert come from this interview.

38. Vanessa Nakate, video, Twitter, January 24 2020, 6:57 a.m., https://twitter.com/vanessa_vash/status/1220722317002756098.
39. Nepeti Nicanor, "Vanessa Nakate's Erasure Portrays an Idealised Climate Activism," *Africa at LSE* (blog), January 31, 2020, https://blogs.lse.ac.uk/africaatlse/2020/01/31/vanessa-nakate-davos-cropped-photo-white-race-climate-activism/.
40. Ashley Lee, "Invisible Networked Publics and Hidden Contention: Youth Activism and Social Media Tactics Under Repression," *New Media & Society* 20, no. 11 (November 2018): 4095–115, https://doi.org/10.1177/1461444818768063.
41. Trump, Thunberg, *MarieClaire*, *Business Insider*, and Ocasio-Cortez quoted in Taylor Telford, "These Self-Described Trolls Tackle Climate Disinformation on Social Media with Wit and Memes," *Washington Post*, July 30, 2021, https://www.washingtonpost.com/business/2021/07/30/greentrolling-big-oil-greenwashing/.
42. Emma Grey Ellis, "Greta Thunberg's Online Attackers Reveal a Grim Pattern," *Wired*, March 4, 2020, https://www.wired.com/story/greta-thunberg-online-harassment/.
43. Ellis, "Greta Thunberg's Online Attackers."
44. Milla Nelson and Meg Vertigan, "Misogyny, Male Rage and the Words Men Use to Describe Greta Thunberg," *The Conversation*, November 12, 2021, http://theconversation.com/misogyny-male-rage-and-the-words-men-use-to-describe-greta-thunberg-124347.
45. Max Hoppenstedt, "Luisa Neubauer über Hatespeech: 'Was über Mich Geschrieben Wird, Ist Schon Krass,'" *Der Spiegel*, August 12, 2020, https://www.spiegel.de/netzwelt/netzpolitik/luisa-neubauer-ueber-hatespeech-was-ueber-mich-geschrieben-wird-ist-schon-krass-a-19bd80bb-4fec-4fea-8e11-85b817c05bf3, my translation.
46. Hirji, "It's Not Just Greta."
47. Jamie Margolin, interviewed by the author, Seattle, October 26, 2019; all quotations from Margolin come from this interview.
48. Hirji, "It's Not Just Greta."
49. Julia Barnett, interviewed by the author, Seattle, October 4, 2019.
50. Rebecca Hersher, "Key Moments in the Dakota Access Pipeline Fight," NPR, February 22, 2017, https://www.npr.org/sections/thetwo-way/2017/02/22/514988040/key-moments-in-the-dakota-access-pipeline-fight.

51. Dina Gilio-Whitaker, *As Long as Grass Grows: The Indigenous Fight for Environmental Justice, from Colonization to Standing Rock* (Boston: Beacon Press, 2019), 12.
52. Jim Gary, "Standing Rock: The Biggest Story That No One's Covering," ZNET, September 11, 2016, https://zcomm.org/znetarticle/standing-rock-the-biggest-story-that-no-ones-covering/.
53. Simon Moya-Smith quoted in Braudie Blais-Billie, "6 Indigenous Activists on Why They're Fighting the Dakota Access Pipeline," *The FADER*, September 9, 2016, https://www.thefader.com/2016/09/09/dakota-access-pipeline-protest-interviews.
54. Tim Giago quoted in Jenni Monet, "Covering Standing Rock," *Columbia Journalism Review*, Spring 2017, https://www.cjr.org/local_news/covering-standing-rock.php.
55. Gaia Casagrande, Mohamed Amine Khaddar, and Stefania Parisi, "Technology and the Local Community: Uses of Drones in #NoDAPL Movement and Dandora Dumpsite Storytelling," *American Behavioral Scientist* 64, no. 13 (November 2020): 1906–20, https://doi.org/10.1177/0002764220952133.
56. Shiyé Bidzííl quoted in "Drone to Be Wild," *Overpass Light Brigade* (blog), n.d., http://overpasslightbrigade.org/drone-to-be-wild/.
57. Jason Koebler, "The Government Is Using a No Fly Zone to Suppress Journalism at Standing Rock," *Vice*, November 30, 2016, https://www.vice.com/en/article/yp3kak/the-government-is-using-a-no-fly-zone-to-suppress-journalism-at-standing-rock.
58. "Divest Your Community," Mazaska Talks, n.d., https://mazaskatalks.org/divest-your-community.
59. Vincent Schilling, "No DAPL Social Media Explosion: Celebrities, Musicians and Politicians Worldwide," *Indian Country Today*, September 11, 2016, https://indiancountrytoday.com/archive/nodapl-social-media-explosion-celebrities-musicians-and-politicians-worldwide; "Stop the Pipelines—No More Tar Sand!," *Lush*, n.d., https://www.lushusa.com/stories/article_ethical-stop-the-pipelines.html.
60. Jeeyun Baik, "The Geotagging Counterpublic: The Case of Facebook Remote Check-Ins to Standing Rock," *International Journal of Communication* 14 (2020): 2057–77.

61. Matt Cagle, "Facebook, Instagram, and Twitter Provided Data Access for a Surveillance Product Marketed to Target Activists of Color," American Civil Liberties Union, October 11, 2016, https://www.aclu.org/news/privacy-technology/facebook-instagram-and-twitter-provided-data-access.
62. Alleen Brown, Will Parrish, and Alice Speri, "Leaked Documents Reveal Counterterrorism Tactics Used at Standing Rock to 'Defeat Pipeline Insurgencies,'" *The Intercept* (blog), May 27, 2017, https://theintercept.com/2017/05/27/leaked-documents-reveal-security-firms-counterterrorism-tactics-at-standing-rock-to-defeat-pipeline-insurgencies/.
63. Document quoted in Casey Newton, "How Facebook Rewards Polarizing Political Ads," *The Verge*, October 11, 2017, https://www.theverge.com/2017/10/11/16449976/facebook-political-ads-trump-russia-election-news-feed.
64. Chris D'Angelo, "Dakota Access Company Bought Up Dozens of Anti-pipeline URLs," *Grist*, April 28, 2019, https://grist.org/article/dakota-access-company-bought-up-dozens-of-anti-pipeline-urls/.
65. Meredith D. Clark, "Shallow Coverage of Standing Rock Is Part of a Bigger Problem," *Poynter* (blog), November 28, 2016, https://www.poynter.org/reporting-editing/2016/shallow-coverage-of-standing-rock-is-part-of-a-bigger-problem/.
66. "Republican Investigation Links Russian Trolls to #NoDAPL Movement," *Indianz*, n.d., https://www.indianz.com/News/2018/03/01/republican-investigation-links-russian-t.asp.
67. Craig Timberg and Tony Romm, "These Provocative Images Show Russian Trolls Sought to Inflame Debate Over Climate Change, Fracking and Dakota Pipeline," *Washington Post*, March 1, 2018, https://www.washingtonpost.com/news/the-switch/wp/2018/03/01/congress-russians-trolls-sought-to-inflame-u-s-debate-on-climate-change-fracking-and-dakota-pipeline/.
68. Matthew Hindman and Vladimir Barash, "Disinformation, 'Fake News' and Influence Campaigns on Twitter," Knight Foundation, October 4, 2018, https://knightfoundation.org/reports/disinformation-fake-news-and-influence-campaigns-on-twitter/.

69. Alleen Brown and Sam Richards, "Pipeline Company Issues Broad Subpoena to News Site That Covered Protests Against It," *The Intercept* (blog), April 3, 2021, https://theintercept.com/2021/04/03/unicorn-riot-dakota-pipeline-energy-transfer-subpoena/.
70. Nina Lakhani, "U.S. Supreme Court Rejects Dakota Access Pipeline Appeal," *The Guardian*, February 22, 2022, https://www.theguardian.com/us-news/2022/feb/22/us-supreme-court-dakota-access-pipeline.

4. COLLECTIVE IMAGINARY

1. Naomi Oreskes and Erik M. Conway, *The Collapse of Western Civilization: A View from the Future* (New York: Columbia University Press, 2014), 2.
2. Jonathan Watts et al., "Half World's Fossil Fuel Assets Could Become Worthless by 2036 in Net Zero Transition," *The Guardian*, November 4, 2021, https://www.theguardian.com/environment/ng-interactive/2021/nov/04/fossil-fuel-assets-worthless-2036-net-zero-transition.
3. Donna Haraway, "A Cyborg Manifesto: Science, Technology, and Socialist-Feminism in the Late Twentieth Century," in *Simians, Cyborgs and Women: The Reinvention of Nature* (New York: Routledge, 1991), 117–58.
4. Coral Davenport and Eric Lipton, "How G.O.P. Leaders Came to View Climate Change as Fake Science," *New York Times*, June 3, 2017, https://www.nytimes.com/2017/06/03/us/politics/republican-leaders-climate-change.html.
5. Emily Grey Ellis, "Greta Thunberg's Online Attackers Reveal a Grim Pattern," *Wired*, March 4, 2020, https://www.wired.com/story/greta-thunberg-online-harassment/.
6. For an overview of the history of denialism and greenwashing in fossil-fuel company advertising, see Geoffery Supran and Naomi Oreskes, "The Forgotten Oil Ads That Told Us Climate Change Was Nothing," *The Guardian*, November 18, 2021, https://www.theguardian.com/environment/2021/nov/18/the-forgotten-oil-ads-that-told-us-climate-change-was-nothing.

7. Ro Khanna quoted in Maxine Joselow and Dino Grandoni, "Big Oil CEOs Testify Before House Oversight Committee," *Washington Post*, October 28, 2021, https://www.washingtonpost.com/climate-enviro nment/2021/10/28/oil-executives-testimony-live-updates/.
8. Rebecca Leber, "What the Oil Industry Still Won't Tell Us," *Vox*, October 28, 2021, https://www.vox.com/22745597/big-oil-congress -hearing-exxonmobil-bp-chevron-shell.
9. Marc Gunther, "Edelman Loses Executives and Clients Over Climate Change Stance," *The Guardian*, July 7, 2015, https://www.theguardian .com/sustainable-business/2015/jul/07/pr-edelman-climate-change -lost-executives-clients.
10. Christine Arena quoted in Adele Peters, "Inside the Campaign to Push PR Firms to Fire Fossil Fuel Clients," *Fast Company*, November 25, 2020, https://www.fastcompany.com/90579261/inside-the -campaign-to-push-pr-firms-to-fire-fossil-fuel-clients.
11. See the CleanCreatives website at https://cleancreatives.org/.
12. Frances Haugen, "Opening Statement to Senate Committee on Commerce, Science & Transportation," *Frances Haugen* (blog), October 5, 2021, https://www.franceshaugen.com/blog/b9xlswihkike7639nn4ie2 30dz9eqy.
13. Alexandra S. Levine et al., "Whistleblower to Senate: Don't Trust Facebook," *Politico*, October 5, 2021, https://www.politico.com/news /2021/10/05/facebook-whistleblower-testifies-congress-515083.
14. Karen Hao, "We Read the Paper That Forced Timnit Gebru out of Google. Here's What It Says," *MIT Technology Review*, December 4, 2020, https://www.technologyreview.com/2020/12/04/1013294/google -ai-ethics-research-paper-forced-out-timnit-gebru/.
15. Cunningham and Costa have since won a settlement requiring Amazon to pay back their wages and post a notice to all employees that they cannot be fired for organizing. See Karen Weise, "Amazon Settles with Activist Workers Who Say They Were Fired," *New York Times*, September 29, 2021, https://www.nytimes.com/2021/09/29/technology /amazon-fired-workers-settlement.html.
16. See Climate Action.tech, "About," n.d., https://climateaction.tech /about/; and Tech Workers Coalition, "Worker Power in the Tech Industry," n.d., https://techworkerscoalition.org/.

17. Amitav Ghosh, *The Great Derangement: Climate Change and the Unthinkable* (Chicago: University of Chicago Press, 2016), 135.
18. Natalie Fenton et al., *The Media Manifesto* (Cambridge: Polity, 2020).
19. Fenton et al., *The Media Manifesto*, 5.
20. Gerald J. Baldasty, *The Commercialization of News in the Nineteenth Century* (Madison: University of Wisconsin Press, 1992).
21. Victor Pickard, *America's Battle for Democracy: The Triumph of Corporate Libertarianism and the Future of Media Reform* (New York: Cambridge University Press, 2015), 126.
22. Martin Moore and Damian Tambini, *Regulating Big Tech: Policy Responses to Digital Dominance* (Oxford: Oxford University Press, 2021), 12.
23. See, for example, Damian Tambini, *Media Freedom* (Cambridge: Polity, 2021); Pickard, *America's Battle for Media Democracy*; Natalie Fenton and Des DJ Freedman, "Media Reform and the Politics of Hope," in *The Future of Media*, ed. Joanna Zylinska, with Goldsmith Media (London: Goldsmith Press, 2022), 25–41; and Tim Wu, *The Curse of Bigness: Antitrust in the New Gilded Age*, Columbia Global Reports (New York: Columbia University, 2018), 75.
24. Arne Hintz, Lina Dencik, and Karin Wahl-Jorgensen, *Digital Citizenship in a Datafied Society* (Cambridge: Polity, 2018).
25. Carole Cadwalladr, "Exposing Cambridge Analytica: It's Been Exhausting, Exhilarating, and Slightly Terrifying," interview by Lee Glendinning, *The Guardian*, September 29, 2018, https://www.theguardian.com/membership/2018/sep/29/cambridge-analytica-cadwalladr-observer-facebook-zuckerberg-wylie.
26. Samantha Bradshaw and Philip Howard, "Troops, Trolls and Troublemakers: A Global Inventory of Organized Social Media Manipulation," Oxford Internet Institute, 2017, https://ora.ox.ac.uk/objects/uuid:cef7e8d9-27bf-4ea5-9fd6-855209b3e1f6.
27. See Shoshana Zuboff, *The Age of Surveillance Capitalism: The Fight for a Human Future at the New Frontier of Power* (London: Profile Books, 2019); and Shoshana Zuboff, "You Are the Object of a Secret Extraction Operation," *New York Times*, November 12, 2021, https://www.nytimes.com/2021/11/12/opinion/facebook-privacy

.html. Also see Nick Couldry and Ulises A. Mejias, *The Costs of Connection: How Data Are Colonizing Human Life and Appropriating It for Capitalism* (Palo Alto, CA: Stanford University Press, 2019).
28. Luke Irwin, "The GDPR: Understanding the 6 Data Protection Principles," *IT Governance European Blog*, December 9, 2021, https://www.itgovernance.eu/blog/en/the-gdpr-understanding-the-6-data-protection-principles.
29. Critics argue that the GDPR and similar regulation actually strengthen the position of Google (Alphabet) and Facebook (Meta) in the digital-advertising market because these large platforms can afford teams of lawyers to ensure compliance with the new complex data requirements. Moreover, because these large platforms offer a wide range of online services to millions of users, they can use the one-time consent given by users for the companies' own benefit. Smaller companies don't have these same resources to deal with the new legislation. So, paradoxically, more regulation (which has to be the same for everyone) and more demands for responsibilities can increase the system's power imbalance.
30. Tim Wu, "Be Afraid of Economic Bigness. Be Very Afraid," *New York Times*, November 11, 2018, https://www.nytimes.com/2018/11/10/opinion/sunday/fascism-economy-monopoly.html?action=click&module=RelatedLinks&pgtype=Article.
31. Wu, *The Curse of Bigness*, 139.
32. Mathias Vermeulen, "Regulating the Digital Public Sphere," Open Society Foundations, June 2021, https://www.opensocietyfoundations.org/uploads/26fac926-c4ce-415e-ae85-a09fcbe10ba4/regulating-the-digital-public-sphere-report-20210617.pdf.
33. Wu, *The Curse of Bigness*, 133.
34. Daisuke Wakabayashi and Tiffany Hsu, "Why Google Backtracked on Its New Search Results Look," *New York Times*, January 31, 2020, https://www.nytimes.com/2020/01/31/technology/google-search-results.html.
35. Frank A. Pasquale, "Internet Nondiscrimination Principles Revisited," *SSRN Electronic Journal*, no. 629 (2020): 1–39, https://doi.org/10.2139/ssm.3634625.

36. Cat Zakrzewski, "Google Calls Itself Green. But It's Still Making Ad Money from Climate-Change Denial," *Washington Post*, December 16, 2021, https://www.washingtonpost.com/technology/2021/12/16/google-climate-change-denial-ads/.
37. Zuboff, *The Age of Surveillance Capitalism*, 128–75.
38. Tonu Basu et al., *Algorithmic Accountability for the Public Sector*, Executive Summary (London: Ada Lovelace Institute, August 24, 2021), https://www.opengovpartnership.org/wp-content/uploads/2021/08/executive-summary-algorithmic-accountability.pdf.
39. European Commission, "The Digital Services Act Package," n.d., https://digital-strategy.ec.europa.eu/en/policies/digital-services-act-package.
40. Victor Pickard, *Democracy Without Journalism? Confronting the Misinformation Society* (Oxford: Oxford University Press, 2019), 5.
41. Victor Pickard, "The Violence of the Market," *Journalism* 20, no. 1 (2019): 156, https://doi.org/10.1177/1464884918808955.
42. Emily Bell, "Do Technology Companies Care About Journalism?," *Columbia Journalism Review*, March 27, 2019, https://www.cjr.org/tow_center/google-facebook-journalism-influence.php.
43. Marc Tracy, "McClatchy, Family-Run News Chain, Goes to Hedge Fund in Bankruptcy Sale," *New York Times*, August 4, 2020, https://www.nytimes.com/2020/08/04/business/media/mcclatchy-newspapers-bankrutpcy-chatham.html.
44. Lucia Walinchus, "We Need a News Utility," *Poynter* (blog), June 16, 2022, https://www.poynter.org/commentary/2022/national-tax-support-journalism/.
45. Cable-Satellite Public Affairs Network (C-SPAN) is a U.S. cable and satellite television network created in 1979 by the cable television industry as a nonprofit public service. It broadcasts many proceedings of the U.S. federal government as well as other public-affairs programming.
46. Pickard, *Democracy Without Journalism?*, 5; on Bell's work, see Ethan Zuckerman, "The Case for Digital Public Infrastructure," Knight First Amendment Institute, Columbia University, 2020, https://knightcolumbia.org/content/the-case-for-digital-public-infrastructure.

47. Katharine Hayhoe, *Saving Us: A Climate Scientist's Case for Hope and Healing in a Divided World* (New York: Simon and Schuster, 2021).
48. Vanessa Bowden, Daniel Nyberg, and Christopher Wright, "'We're Going Under': The Role of Local News Media in Dislocating Climate Change Adaptation," *Environmental Communication* 15, no. 5 (2021): 625–40.
49. Candis Callison and Mary Lynn Young, *Reckoning: Journalism's Limits and Possibilities* (Oxford: Oxford University Press, 2019).
50. Allissa V. Richardson, *Bearing Witness While Black: African Americans, Smartphones, and the New Protest #Journalism* (New York: Oxford University Press, 2020); Catherine R. Squires, "Rethinking the Black Public Sphere: An Alternative Vocabulary for Multiple Public Spheres," *Communication Theory* 12, no. 4 (2002): 446–68.
51. Callison and Young, *Reckoning*, 203.
52. Matt Carlson, *Journalistic Authority: Legitimating News in the Digital Era* (New York: Columbia University Press, 2017).
53. Risto Kunelius, "A Forced Opportunity: Climate Change and Journalism," *Journalism* 20, no. 1 (2019): 218–21, https://doi.org/10.1177/1464884918807596.
54. Fenton et al., *The Media Manifesto*, 95–96.
55. Barbie Zelizer, C. W. Anderson, and Pablo J. Boczkowski, *The Journalism Manifesto* (Cambridge: Polity, 2022), 19.
56. Callison and Young, *Reckoning*; Adrienne Russell, *Journalism as Activism: Recoding Media Power* (Cambridge: Polity, 2016).
57. Richardson, *Bearing Witness While Black*.
58. Andrew Chadwick, *The Hybrid Media System: Politics and Power* (New York: Oxford University Press, 2013), 208.
59. Adrienne Russell, "Coming to Terms with Dysfunctional Hybridity: A Conversation with Andrew Chadwick on the Challenges to Liberal Democracy in the Second-Wave Networked Era," *Studies in Communication Sciences* 20, no. 2 (2020): 211–25.
60. For one example of how the business of journalism threatens democracy, see Jacey Fortin and Jonah Engel Bromwich, "Sinclair Made Dozens of Local News Anchors Recite the Same Script," *New York Times*, April 2, 2018, https://www.nytimes.com/2018/04/02/business/media/sinclair-news-anchors-script.html.

61. Charles Duhigg, "What Does Your Credit-Card Company Know About You?," *New York Times*, May 12, 2009, https://www.nytimes.com/2009/05/17/magazine/17credit-t.html?pagewanted=all.
62. See Julia Angwin and Terry Parris Jr., "Facebook Lets Advertisers Exclude Users by Race," *ProPublica*, October 28, 2016, https://www.propublica.org/article/facebook-lets-advertisers-exclude-users-by-race; Jennifer Valentino-DeVries, Jeremy Singer-Vine, and Ashkan Soltani, "Websites Vary Prices, Deals Based on Users' Information," *Wall Street Journal*, December, 24, 2012, https://www.wsj.com/articles/SB10001424127887323777204578189391813881534; and Rachel Treisman, "Facebook Fell Short of Its Promises to Label Climate Change Denial, a Study Finds," NPR, February 23, 2022, https://www.npr.org/2022/02/23/1082561725/facebook-climate-change-label.
63. Zizi Papacharissi, *After Democracy: Imagining Our Political Future* (New Haven, CT: Yale University Press, 2021), 127.
64. Dipesh Chakrabarty, "The Climate of History: Four Theses," *Critical Inquiry* 35, no. 2 (2009): 208, https://doi.org/10.1086/596640.
65. See, for example, Haraway, "A Cyborg Manifesto"; Christine Harold, *Things Worth Keeping: The Value of Attachment in a Disposable World* (Minneapolis: University of Minnesota Press, 2020); and Bruno Latour, *Reassembling the Social: An Introduction to Actor-Network-Theory* (Oxford: Oxford University Press, 2005).
66. See, for example, Enrique Salmón, "Kincentric Ecology: Indigenous Perceptions of the Human–Nature Relationship," *Ecological Applications* 10, no. 5 (2000): 1327–32, https://doi.org/10.1890/1051-0761(2000)010[1327:KEIPOT]2.0.CO;2; Vanessa Watts, "Indigenous Place–Thought and Agency Amongst Humans and Non Humans," *Decolonization: Indigeneity, Education & Society* 2, no. 1 (2013): 20–34, https://jps.library.utoronto.ca/index.php/des/article/view/19145; and Dina Gilio-Whitaker, *As Long as Grass Grows: The Indigenous Fight for Environmental Justice, from Colonization to Standing Rock* (Boston: Beacon Press, 2019).
67. See, for example, Arizona State University Institute for Humanities Research, "Black Ecologies Initiative," n.d., https://ihr.asu.edu/black-ecologies.
68. Jane Bennett, *Vibrant Matter: A Political Ecology of Things* (Durham, NC: Duke University Press, 2010).

69. Whitney Phillips and Ryan M. Milner, *You Are Here: A Field Guide for Navigating Polarized Speech, Conspiracy Theories, and Our Polluted Media Landscape* (Cambridge, MA: MIT Press, 2021), 8.
70. Dipesh Chakrabarty, "The Climate of History: Four Theses," in *Energy Humanities: An Anthology*, ed. Imre Szeman and Dominic Boyer (Baltimore, MD: Johns Hopkins University Press, 2017), 40.
71. Robert Jay Lifton, quoted in Audrea Lim, "The Ideology of Fossil Fuels," *Dissent*, Spring 2018, https://www.dissentmagazine.org/article/ideology-fossil-fuels-apocalypse-petrocapitalism-energy-humanities. See also Robert Jay Lifton, *The Climate Swerve: Reflections on Mind, Hope, and Survival* (New York: New Press, 2017).
72. Chris Hayes, "The New Abolitionism," *The Nation*, April 22, 2014, https://www.thenation.com/article/archive/new-abolitionism/.
73. Robert Bullard, "Addressing Environmental Racism" (interview), *Journal of International Affairs* 73, no. 1 (2019), https://jia.sipa.columbia.edu/addressing-environmental-racism.
74. Lifton quoted in Lim, "The Ideology of Fossil Fuels."
75. "1485 Institutions with Assets Over $39.2 Trillion Have Committed to Divest from Fossil Fuels," Stand.earth, October 26, 2021, https://www.stand.earth/divestinvest2021.
76. Bill McKibben, "This Movement Is Taking Money Away from Fossil Fuels, and It's Working," *New York Times*, October 26, 2021, https://www.nytimes.com/2021/10/26/opinion/climate-change-divestment-fossil-fuels.html.
77. Jennifer Hiller and Svea Herbst-Bayliss, "Exxon Loses Board Seats to Activist Hedge Fund in Landmark Climate Vote," Reuters, May 26, 2021, https://www.reuters.com/business/sustainable-business/shareholder-activism-reaches-milestone-exxon-board-vote-nears-end-2021-05-26/.
78. Stanley Reed and Claire Moses, "A Dutch Court Rules That Shell Must Step Up Its Climate Change Efforts," *New York Times*, October 28, 2021, https://www.nytimes.com/2021/05/26/business/royal-dutch-shell-climate-change.html.
79. For a full list of divestment actions against fossil-fuel companies, refer to the Global Fossil Fuel Divestment Commitments Database at https://divestmentdatabase.org/.

80. For more detail on the Inflation Reduction Act climate-justice provision, see Environmental and Energy Law Program, Harvard University, "Environmental Justice Provisions of the 2022 Inflation Reduction Act," table, n.d., http://eelp.law.harvard.edu/wp-content/uploads/EELP-IRA-EJ-Provisions-Table.pdf.

BIBLIOGRAPHY

Abernathy, Penny. "The State of Local News: The 2022 Report." Local News Initiative, June 29, 2022. https://localnewsinitiative.northwestern.edu/research/state-of-local-news/report/.

Akin, Heather, and Dietram A. Scheufele. "Overview of the Science of Science Communication." In *The Oxford Handbook of the Science of Science Communication*, ed. Katherine Hall Jamieson, Dan Kahan, and Dietram A. Scheufele, 25–33. New York: Oxford University Press, 2017.

Allgaier, Joachim. "Science and Environmental Communication on YouTube: Strategically Distorted Communications in Online Videos on Climate Change and Climate Engineering." *Frontiers in Communication* 4 (2019). https://doi.org/10.3389/fcomm.2019.00036.

Ananny, Mike. *Networked Press Freedom: Creating Infrastructures for a Public Right to Hear*. Cambridge, MA: MIT Press, 2018.

——. "Tech Platforms Are Where Public Life Is Increasingly Constructed, and Their Motivations Are Far from Neutral." NiemanLab blog, October 10, 2019. https://www.niemanlab.org/2019/10/tech-platforms-are-where-public-life-is-increasingly-constructed-and-their-motivations-are-far-from-neutral/.

Anderson, Ashley A., Dominique Brossard, Dietram A. Scheufele, Michael A. Xenos, and Peter Ladwig. "The 'Nasty Effect': Online Incivility and Risk Perceptions of Emerging Technologies." *Journal of Computer-Mediated Communication* 19, no. 3 (2014): 373–87. http://doi.org/10.1111/jcc4.12009.

Anderson, Benedict. *Imagined Communities: Reflections on the Origin and Spread of Nationalism*. London: Verso, 1983.

Anderson, Chris. "The End of Theory: The Data Deluge Makes the Scientific Method Obsolete." *Wired*, June 23, 2008. https://www.wired.com/2008/06/pb-theory/.

Andı, Simge. "How People Access News About Climate Change." *Digital News Report*, March 2020. https://www.digitalnewsreport.org/survey/2020/how-people-access-news-about-climate-change.

Angwin, Julia, Jeff Larson, Surya Mattu, and Lauren Kirchner. "Machine Bias." *ProPublica*, May 23, 2016. https://www.propublica.org/article/machine-bias-risk-assessments-in-criminal-sentencing.

Angwin, Julia, and Terry Parris Jr. "Facebook Lets Advertisers Exclude Users by Race." *ProPublica*, October 28, 2016. https://www.propublica.org/article/facebook-lets-advertisers-exclude-users-by-race.

Ardia, David S., Evan Ringel, Victoria Ekstrand, and Ashley Fox. "Addressing the Decline of Local News, Rise of Platforms, and Spread of Mis- and Disinformation Online: A Summary of Current Research and Policy Proposals." University of North Carolina Legal Studies Research Paper, 2020. https://citap.unc.edu/local-news-platforms-mis-disinformation/.

Arendt, Hannah. *The Human Condition*. 1958. Reprint. Chicago: University of Chicago Press, 1998.

Arizona State University Institute for Humanities Research. "Black Ecologies Initiative." N.d. https://ihr.asu.edu/black-ecologies.

Aronczyk, Melissa. "Public Communication in a Promotional Culture." In *Rethinking Media Research for Changing Societies*, ed. Matthew Powers and Adrienne Russell, 39–50. Cambridge: Cambridge University Press, 2020.

Aronczyk, Melissa, and Maria I. Espinoza. *A Strategic Nature: Public Relations and the Politics of American Environmentalism*. Oxford: Oxford University Press, 2021.

Aronoff, Kate. "Things Are Bleak! Jonathan Safran Foer's Quest for Planetary Salvation." *The Nation*, October 29, 2019. https://www.thenation.com/article/archive/jonathan-safran-foer-we-are-the-weather-climate-review.

Åström, Victor. "Environmental Report Defenders Under Attack." Swedish Society for Nature Conservation, November 2019. https://www.universal-rights.org/wp-content/uploads/2019/11/environmental_defenders_under_attack_eng.pdf.

Bagley, Katherine, and Naveena Sadasivam. "Climate Denial's Ugly Side: Hate Mail to Scientists." *Inside Climate News*, December 11, 2015. https://insideclimatenews.org/news/11122015/climate-change-global-warming-denial-ugly-side-scientists-hate-mail-hayhoe-mann.

Baik, Jeeyun. "The Geotagging Counterpublic: The Case of Facebook Remote Check-Ins to Standing Rock." *International Journal of Communication* 14 (2020): 2057–77.

Bail, Chris. *Breaking the Social Media Prism*. Princeton, NJ: Princeton University Press, 2021.

Baldasty, Gerald J. *The Commercialization of News in the Nineteenth Century*. Madison: University of Wisconsin Press, 1992.

Ballew, Matthew, Edward Maibach, John Kotcher, Parrish Bergquist, Seth Rosenthal, Jennifer Marion, and Anthony Lieserowitz. "Which Racial/Ethnic Groups Care Most About Climate Change?" Yale Program on Climate Change Communication blog, April 16, 2020. https://climatecommunication.yale.edu/publications/race-and-climate-change/.

Barassi, Veronica. *Activism on the Web: Everyday Struggles Against Digital Capitalism*. London: Routledge, 2015.

Barbaro, Michael. "The Big Tech Hearing." *The Daily* (podcast), *New York Times*, July 30, 2020. https://www.nytimes.com/2020/07/30/podcasts/the-daily/congress-facebook-amazon-google-apple.html.

——. "Jack Dorsey on Twitter's Mistakes." *The Daily* (podcast), *New York Times*, August 7, 2020. https://www.nytimes.com/2020/08/07/podcasts/the-daily/Jack-dorsey-twitter-trump.html.

Barlow, John Perry. "A Declaration of the Independence of Cyberspace" (1996). *Duke Law & Technology Review* 18, no. 1 (2019): 5–7.

Basu, Tonu, Jenny Brennan, Amba Kak, and Divij Joshi. *Algorithmic Accountability for the Public Sector*. Executive Summary. London: Ada Lovelace Institute, August 24, 2021. https://www.opengovpartnership.org/wp-content/uploads/2021/08/executive-summary-algorithmic-accountability.pdf.

BBC News. "Catherine McKenna: Canada Environment Minister Given Extra Security." September 8, 2019. https://www.bbc.com/news/world-us-canada-49627153.

Bell, Emily. "Do Technology Companies Care About Journalism?" *Columbia Journalism Review*, March 27, 2019. https://www.cjr.org/tow_center/google-facebook-journalism-influence.php.

Bell, Emily J., Taylor Owen, Peter D. Brown, Codi Hauka, and Nushin Rashidian. "The Platform Press: How Silicon Valley Reengineered Journalism." Tow Center for Digital Journalism, March 29, 2017. https://www.cjr.org/tow_center_reports/platform-press-how-silicon-valley-reengineered-journalism.php.

Bell James, Jacob Poushter, Moira Fagan, and Christine Huang. "In Response to Climate Change, Citizens in Advanced Economies Are Willing to Alter How They Live and Work." Pew Research Center, September 14, 2021. https://www.pewresearch.org/global/2021/09/14/in-response-to-climate-change-citizens-in-advanced-economies-are-willing-to-alter-how-they-live-and-work/.

Benkler, Yochai. *The Wealth of Networks: How Social Production Transforms Markets and Freedom*. New Haven, CT: Yale University Press, 2006.

Benkler, Yochai, Robert Faris, and Hal Roberts. *Network Propaganda: Manipulation, Disinformation, and Radicalization in American Politics*. Oxford: Oxford Scholarship Online, 2018.

Bennett, Jane. *Vibrant Matter: A Political Ecology of Things*. Durham, NC: Duke University Press, 2010.

Bennett, W. Lance, and Barbara Pfetsch. "Rethinking Political Communication in a Time of Disrupted Public Spheres." *Journal of Communication* 68, no. 2 (2018): 243–53. https://doi.org/10.1093/joc/jqx017.

Bennett, W. Lance, and Steven Livingston, eds. *The Disinformation Age: Politics, Technology and Disruptive Communication in the United States*. Cambridge: Cambridge University Press, 2021.

———. "A Brief History of the Disinformation Age: Information Wars and the Decline of Institutional Authority." In *The Disinformation Age: Politics, Technology, and Disruptive Communication in the United States*, ed. W. Lance Bennett and Steven Livingston, 3–40. Cambridge: Cambridge University Press, 2021.

Bimber, Bruce, and Homero Gil de Zúñiga. "The Unedited Public Sphere." *New Media & Society* 22, no. 4 (2020): 700–715. https://doi.org/10.1177/1461444819893980.

Bitar, Jenna. "6 Ways Government Is Going After Environmental Activists." *ACLU News*, February 6, 2018. https://www.aclu.org/news/free-speech/6-ways-government-going-after-environmental-activists.

Black, James F. "James Black Talk (1977)." *Inside Climate News*, September 15, 2015. https://insideclimatenews.org/documents/james-black-1977-prese ntation.

Blais-Billie, Braudie. "6 Indigenous Activists on Why They're Fighting the Dakota Access Pipeline." *The FADER*, September 9, 2016. https:// www.thefader.com/2016/09/09/dakota-access-pipeline-protest-inter views.

Bowden, Vanessa, Daniel Nyberg, and Christopher Wright. "'We're Going Under': The Role of Local News Media in Dislocating Climate Change Adaptation." *Environmental Communication* 15, no. 5 (2021): 625–40.

Bowker, Geoffrey C., and Susan Leigh Star. *Sorting Things Out: Classification and Its Consequences*. Cambridge, MA: MIT Press, 2000.

boyd, danah. "Media Manipulation, Strategic Amplification, and Responsible Journalism." *Data & Society: Points*, September 14, 2018. https:// points.datasociety.net/media-manipulation-strategic-amplification -and-responsible-journalism-95f4d611f462.

Boykoff, Maxwell T. *Creative (Climate) Communications: Productive Pathways for Science, Policy and Society*. Cambridge: Cambridge University Press, 2019.

———. *Who Speaks for the Climate? Making Sense of Media Reporting on Climate Change*. Cambridge: Cambridge University Press, 2011.

Boykoff, Maxwell T., and Jules M. Boykoff. "Balance as Bias: Global Warming and the U.S. Prestige Press." *Global Environmental Change* 14, no. 2 (2004): 125–36.

———. "Climate Change and Journalistic Norms: A Case-Study of U.S. Mass-Media Coverage." *Geoforum* 38, no. 6 (2007): 1190–204. https://doi .org/10.1016/j.geoforum.2007.01.008.

Bradshaw, Samantha, and Philip Howard. "Troops, Trolls and Troublemakers: A Global Inventory of Organized Social Media Manipulation." Oxford Internet Institute, 2017. https://ora.ox.ac.uk/objects/uuid:cef7e8d9 -27bf-4ea5-9fd6-855209b3e1f6.

Brenan, Megan. "Media Confidence Ratings at Record Lows." Gallup, July 18, 2022. https://news.gallup.com/poll/394817/media-confidence -ratings-record-lows.aspx.

Brooks, Ryan. "How Russians Attempted to Use Instagram to Influence Native Americans." *Buzzfeed*, October 23, 2017. https://www.buzzfeednews

.com/article/ryancbrooks/russian-troll-efforts-extended-to-standing-rock#.fe1JrJXRJ.

Brossard, Dominique, and Dietram A. Scheufele. "Science, New Media, and the Public." *AAAS*, January 4, 2013. http://www.sciencemag.org/content/339/6115/40.short.

Brown, Alleen, Will Parrish, and Alice Speri. "Leaked Documents Reveal Counterterrorism Tactics Used at Standing Rock to 'Defeat Pipeline Insurgencies.'" *The Intercept* (blog), May 27, 2017. https://theintercept.com/2017/05/27/leaked-documents-reveal-security-firms-counterterrorism-tactics-at-standing-rock-to-defeat-pipeline-insurgencies/.

Brown, Alleen, and Sam Richards. "Pipeline Company Issues Broad Subpoena to News Site That Covered Protests Against It." *The Intercept* (blog), April 3, 2021. https://theintercept.com/2021/04/03/unicorn-riot-dakota-pipeline-energy-transfer-subpoena/.

Brüggemann, Michael. "Post-normal Journalism: Climate Journalism and Its Changing Contribution to an Unsustainable Debate." In *What Is Sustainable Journalism? Integrating the Environmental, Social, and Economic Challenges of Journalism*, ed. Peter Berglez, Ulrika Olausson, and Mart Ots, 57–73. New York: Peter Lang, 2017.

Brüggemann, Michael, and Sven Engesser. "Between Consensus and Denial: Climate Journalists as Interpretive Community." *Science Communication* 36, no. 4 (2014): 399–427. https://doi.org/10.1177/1075547014533662.

Brulle, Robert J. "Institutionalizing Delay: Foundation Funding and the Creation of U.S. Climate Change Counter-Movement Organizations." *Climatic Change* 122, no. 4 (2014): 681–94. https://doi.org/10.1007/s10584-013-1018-7.

Brünker, Felix, Fabian Deitelhoff, and Milad Mirbabaie. "Collective Identity Formation on Instagram—Investigating the Social Movement Fridays for Future." Paper presented at the Australasian Conference on Information Systems, Perth, December 11, 2019. https://arxiv.org/abs/1912.05123.

Bruns, Axel. *Blogs, Wikipedia, Second Life, and Beyond: From Production to Produsage*. Digital Formations, vol. 45. New York: Peter Lang, 2008.

Bullard, Robert. "Addressing Environmental Racism" (interview). *Journal of International Affairs* 73, no. 1 (2019). https://jia.sipa.columbia.edu

/addressing-environmental-racism.Bullard, Robert, Maaz Gardezi, Carrie Chennault, and Hannah Dankbar. "Climate Change and Environmental Justice: A Conversation with Dr. Robert Bullard." *Journal of Critical Thought and Praxis* 5, no. 2 (2016). https://iastatedigitalpress.com/jctp/article/566/galley/446/view/.

Bursztynsky, Jessica. "Vice Media CEO Slams Big Tech as 'Great Threat to Journalism' in Layoffs Memo." CNBC, May 15, 2020. https://www.cnbc.com/2020/05/15/vice-media-ceo-slams-big-tech-as-great-threat-to-journalism.html.

Cadwalladr, Carole. "Exposing Cambridge Analytica: It's Been Exhausting, Exhilarating, and Slightly Terrifying." Interview by Lee Glendinning. *The Guardian*, September 29, 2018. https://www.theguardian.com/membership/2018/sep/29/cambridge-analytica-cadwalladr-observer-facebook-zuckerberg-wylie.

Cadwalladr, Carole, and Emma Graham-Harrison. "Revealed: 50 Million Facebook Profiles Harvested for Cambridge Analytica in Major Data Breach." *The Guardian*, March 17, 2018. https://www.theguardian.com/news/2018/mar/17/cambridge-analytica-facebook-influence-us-election.

Cagle, Matt. "Facebook, Instagram, and Twitter Provided Data Access for a Surveillance Product Marketed to Target Activists of Color." American Civil Liberties Union, October 11, 2016. https://www.aclu.org/news/privacy-technology/facebook-instagram-and-twitter-provided-data-access.

Calhoun, Craig. "Facets of the Public Sphere: Dewey, Arendt, Habermas." In *Institutional Change in the Public Sphere: Views on the Nordic Model*, ed. Fredrik Engelstad, Håkon Larsen, Jon Rogstad, Kari Steen-Johnsen, Dominika Polkowska, Andrea S. Dauber-Griffin, and Adam Leverton, 23–45. Berlin: De Gruyter, 2017.

Callison, Candis. *How Climate Change Comes to Matter: The Communal Life of Facts*. Durham, NC: Duke University Press, 2014.

Callison, Candis, and Mary Lynn Young. *Reckoning: Journalism's Limits and Possibilities*. Oxford: Oxford University Press, 2019.

Calma, Justine. "Chan Zuckerberg Initiative Announces Tens of Millions in Funding for Climate Tech." *The Verge*, February 10, 2022. https://www.theverge.com/2022/2/10/22927245/chan-zuckerberg-initiative-millions-funding-climate-tech-carbon-removal.

Cammaerts, Bart. "The Mediated Circulation of the UK YouthStrike4Climate Movement's Discourses and Actions." *European Journal of Cultural Studies* (forthcoming).

———. "Protest Logics and the Mediation Opportunity Structure." *European Journal of Communication* 27, no. 2 (June 2012): 117–34. https://doi.org/10.1177/0267323112441007.

Carbon Brief. "Exclusive: BBC Issues Internal Guidance on How to Report Climate Change." N.d. https://www.carbonbrief.org/exclusive-bbc-issues-internal-guidance-on-how-to-report-climate-change.

Carlson, Matt. *Journalistic Authority: Legitimating News in the Digital Era.* New York: Columbia University Press, 2017.

Carrington, Damian. "It's Over for Fossil Fuels: IPCC Spells Out What's Needed to Avert Climate Disaster." *The Guardian*, April 4, 2022. https://www.theguardian.com/environment/2022/apr/04/its-over-for-fossil-fuels-ipcc-spells-out-whats-needed-to-avert-climate-disaster.

———. "Why the *Guardian* Is Changing the Language It Uses About the Environment." *The Guardian*, May 17, 2019. https://www.theguardian.com/environment/2019/may/17/why-the-guardian-is-changing-the-language-it-uses-about-the-environment.

Casagrande, Gaia, Mohamed Amine Khaddar, and Stefania Parisi. "Technology and the Local Community: Uses of Drones in #NoDAPL Movement and Dandora Dumpsite Storytelling." *American Behavioral Scientist* 64, no. 13 (November 2020): 1906–20. https://doi.org/10.1177/0002764220952133.

Castells, Manuel. *Communication Power.* New York: Oxford University Press, 2009.

———. "An Introduction to the Information Age." *Colonial Latin American Review* 2, no. 7 (1997): 6–16. https://doi.org/10.1080/13604819708900050.

Ceccarelli, Leah. "The Defense of Science in the Public Sphere." Paper presented at the International Society for the Study of Argumentation conference, Amsterdam, July 3–6, 2018.

Chadwick, Andrew. *The Hybrid Media System: Politics and Power.* New York: Oxford University Press, 2013.

———. *The Hybrid Media System: Politics and Power.* 2nd ed. New York: Oxford University Press, 2017.

Chakrabarty, Dipesh. "The Climate of History: Four Theses." *Critical Inquiry* 35, no. 2 (2009): 197–222. https://doi.org/10.1086/596640.
——. "The Climate of History: Four Theses." In *Energy Humanities: An Anthology*, ed. Imre Szeman and Dominic Boyer, 32–54. Baltimore, MD: Johns Hopkins University Press, 2017.
Chan, Sewall. "Since 2005, Texas Has Lost More Newspaper Journalists per Capita Than All but Two Other States." *Texas Tribune*, June 29, 2022. https://www.texastribune.org/2022/06/29/death-local-news-texas/.
Clark, Meredith D. "Shallow Coverage of Standing Rock Is Part of a Bigger Problem." *Poynter* (blog), November 28, 2016. https://www.poynter.org/reporting-editing/2016/shallow-coverage-of-standing-rock-is-part-of-a-bigger-problem/.
Cleaver, Harry. "The Zapatistas and the Electronic Fabric of Struggle." In *Zapatista! Reinventing Revolution in Mexico*, ed. John Holloway and Eloina Pelaez, 621–40. Chicago: Pluto, 1998.
Cohen, Bernard C. *The Press and Foreign Policy*. Princeton, NJ: Princeton University Press, 1963.
Cooper, Evlondo. "How Broadcast TV News Covered Environmental Justice in 2021." Media Matters, May 10, 2022. https://www.mediamatters.org/broadcast-networks/how-broadcast-tv-news-covered-environmental-justice-2021.
Cooper, Evlondo, and Allison Fisher. "Covering Climate Now's Innovative Model Sets a New Standard for More and Better Climate Coverage." Media Matters, November 5, 2019. https://www.mediamatters.org/broadcast-networks/covering-climate-nows-innovative-model-sets-new-standard-more-and-better-climate.
Couldry, Nick. "The Myth of 'Us': Digital Networks, Political Change and the Production of Collectivity." *Information, Communication & Society* 18, no. 6 (2015): 608–26.
Couldry, Nick, and Ulises A. Mejias. *The Costs of Connection: How Data Are Colonizing Human Life and Appropriating It for Capitalism*. Palo Alto, CA: Stanford University Press, 2019.
Craft, Stephanie, and Morten S. Kristensen. "Noise and the Values of News." In *Rethinking Media Research for Changing Societies*, ed. Matthew Powers and Adrienne Russell, 78–88. Cambridge: Cambridge University Press, 2020.

Crawford, Kate. *The Atlas of AI: Power, Politics, and the Planetary Costs of Artificial Intelligence*. New Haven, CT: Yale University Press, 2021.

Dahlgren, Peter. "Media, Knowledge and Trust: The Deepening Epistemic Crisis of Democracy." *Javnost—The Public* 25, nos. 1–2 (2018): 20–27.

D'Angelo, Chris. "Dakota Access Company Bought Up Dozens of Anti-pipeline URLs." *Grist*, April 28, 2019. https://grist.org/article/dakota-access-company-bought-up-dozens-of-anti-pipeline-urls/.

Davenport, Coral, and Eric Lipton. "How G.O.P. Leaders Came to View Climate Change as Fake Science." *New York Times*, June 3, 2017. https://www.nytimes.com/2017/06/03/us/politics/republican-leaders-climate-change.html.

Democracia Abierta. "Democracy, Authoritarianism and the Climate Emergency." *openDemocracy*, September 18, 2019. https://www.opendemocracy.net/en/democraciaabierta/democracia-autoritarismo-y-emergencia-clim%C3%A1tica-en/.

Dencik, Lina, Arne Hintz, and Jonathan Cable. "Towards Data Justice? The Ambiguity of Anti-surveillance Resistance in Political Activism." *Big Data & Society* 3, no. 2 (December 2016): 1–12. https://doi.org/10.1177/2053951716679678.

Dewey, John. *The Public and Its Problems*. 1927. Edited by Melvin L. Rogers. Athens, GA: Swallow Press, 2016.

Dickson, David. "The Case for a 'Deficit Model' of Science Communication." *Science and Development Network*, June 24, 2005. https://www.scidev.net/global/editorials/the-case-for-a-deficit-model-of-science-communic/.

Digital News Report 2022. Oxford: Reuters Institute for the Study of Journalism, June 15, 2022. https://reutersinstitute.politics.ox.ac.uk/digital-news-report/2022.

Duhigg, Charles. "What Does Your Credit-Card Company Know About You?" *New York Times*, May 12, 2009. https://www.nytimes.com/2009/05/17/magazine/17credit-t.html?pagewanted=all.

Eide, Elizabeth, and Risto Kunelius, eds. *Media Meets Climate: The Global Challenge for Journalism*. Gothenberg, Sweden: Nordicom, University of Gothenburg, 2012.

Ellis, Emily Grey. "Greta Thunberg's Online Attackers Reveal a Grim Pattern." *Wired*, March 4, 2020. https://www.wired.com/story/greta-thunberg-online-harassment/.

Entman, Robert M. "Framing: Toward Clarification of a Fractured Paradigm." *Journal of Communication* 43, no. 4 (1993): 51–58.

Environmental and Energy Law Program, Harvard University. "Environmental Justice Provisions of the 2022 Inflation Reduction Act." Table, n.d. http://eelp.law.harvard.edu/wp-content/uploads/EELP-IRA-EJ-Provisions-Table.pdf.

Erviti, María Carmen, José Azevedo, and Mónica Codina. "When Science Becomes Controversial." In *Communicating Science and Technology Through Online Video*, ed. Bienvenido León and Michael Bourk, 41–54. London: Routledge, 2018.

Escobar, Herton. "Brazil's New President Has Scientists Worried. Here's Why." *Science*, January 22, 2019. https://www.science.org/content/article/brazil-s-new-president-has-scientists-worried-here-s-why.

European Commission. "The Digital Services Act Package." N.d. https://digital-strategy.ec.europa.eu/en/policies/digital-services-act-package.

Extinction Rebellion. "XR NYC STANDARDS FOR MEDIA." N.d. [c. June 2019]. https://www.xrebellion.nyc/media-standards.

Farrell, Justin, Kathryn McConnell, and Robert Brulle. "Evidence-Based Strategies for Combat Scientific Misinformation." *Nature Climate Change* 9 (2019): 191–95. https://doi.org/10.1038/s41558-018-0368-6.

FBI Portland. "Statement on Misinformation Related to Wildfires." September 11, 2020. https://www.fbi.gov/contact-us/field-offices/portland/news/press-releases/fbi-releases-statement-on-misinformation-related-to-wildfires.

Fenton, Natalie, and Des DJ Freedman. "Media Reform and the Politics of Hope." In *The Future of Media*, ed. Joanna Zylinska, with Goldsmith Media, 25–41. London: Goldsmith Press, 2022.

Fenton, Natalie, Des Freeman, Justin Schlosberg, and Lina Dencik. *The Media Manifesto*. Cambridge: Polity, 2020.

Fitri, Afiq. "The Tech Industry's Progress on Carbon Emissions Has Been Mixed." *TechMonitor*, June 23, 2022. https://techmonitor.ai/leadership/sustainability/tech-industry-carbon-emissions-progress.

Fortin, Jacey, and Jonah Engel Bromwich. "Sinclair Made Dozens of Local News Anchors Recite the Same Script." *New York Times*, April 2, 2018. https://www.nytimes.com/2018/04/02/business/media/sinclair-news-anchors-script.html.

Francisco, Sara C., and Diane H. Felmlee. "What Did You Call Me? An Analysis of Online Harassment Towards Black and Latinx Women." *Race and Social Problems* 14, no. 1 (2022): 1–13.

Fraser, Nancy. "Rethinking the Public Sphere: A Contribution to the Critique of Actually Existing Democracy." *Social Text*, nos. 25–26 (1990): 56–80.

——. *Scales of Justice: Reimagining Political Space in a Globalizing World*. New York: Columbia University Press, 2018.

Freedman, Eric. "Environmental Journalists Face Threats, Violence." International Journalists Network, November 30, 2018. https://ijnet.org/en/story/environmental-journalists-face-threats-violence.

Fridays for Future. "Reasons to Strike." Blog, n.d. https://fridaysforfuture.org/take-action/reasons-to-strike/.

Funk, Cary. "Key Findings: How Americans' Attitudes About Climate Change Differ by Generation, Party and Other Factors." Pew Research Center blog, May 26, 2021. https://www.pewresearch.org/fact-tank/2021/05/26/key-findings-how-americans-attitudes-about-climate-change-differ-by-generation-party-and-other-factors/.

Gary, Jim. "Standing Rock: The Biggest Story That No One's Covering." ZNET, September 11, 2016. https://zcomm.org/znetarticle/standing-rock-the-biggest-story-that-no-ones-covering/.

Ghosh, Amitav. *The Great Derangement: Climate Change and the Unthinkable*. Chicago: University of Chicago Press, 2016.

——. *The Nutmeg's Curse: Parables for a Planet in Crisis*. Chicago: University of Chicago Press, 2021.

Gifford, Robert. "The Dragons of Inaction: Psychological Barriers That Limit Climate Change Mitigation and Adaptation." *American Psychologist* 66, no. 4 (2011): 290–302. https://doi.org/10.1037/a0023566.

Gilio-Whitaker, Dina. *As Long as Grass Grows: The Indigenous Fight for Environmental Justice, from Colonization to Standing Rock*. Boston: Beacon Press, 2019.

Gillespie, Tarleton. "Governance of and by Platforms." In *The SAGE Handbook of Social Media*, ed. Jean Burgess, Thomas Poell, and Alice Marwick, 254–78. Los Angeles: SAGE, 2017.

Gitlin, Todd. *The Whole World Is Watching*. Berkeley: University of California Press, 1980.

Glasser, Theodore. "Objectivity Precludes Responsibility." *The Quill* 72, no. 2 (1984): 13–16.

———. "Public Journalism Movement." In *The International Encyclopedia of Political Communication*, vol. 3, ed. Gianpietro Mazzoleni. Chichester, U.K.: Wiley Blackwell, 2015. https://onlinelibrary.wiley.com/doi/10.1002/9781118541555.wbiepc203.

Goffman, Erving. *Frame Analysis: An Essay on the Organization of Experience*. Cambridge, MA: Harvard University Press, 1974.

Gold, Ashley. "Big Tech Antitrust Hearing." *Washington Journal*, C-SPAN, July 29, 2020. https://www.c-span.org/video/?474239-101/washington-journal-ashley-gold-big-tech-antitrust-hearing.

Griffin, Paul. *The Carbon Majors Database: CDP Carbon Majors Report 2017*. Snowmass, CO: CDP, Climate Accountability Institute, July 2017. https://cdn.cdp.net/cdp-production/cms/reports/documents/000/002/327/original/Carbon-Majors-Report-2017.pdf?1501833772.

Griswold, Eliza. "How 'Silent Spring' Ignited the Environmental Movement." *New York Times Magazine*, September 23, 2012. https://www.nytimes.com/2012/09/23/magazine/how-silent-spring-ignited-the-environmental-movement.html.

Guber, Deborah Lynn, Jeremiah Bohr, and Riley E. Dunlap. "'TIME TO WAKE UP': Climate Change Advocacy in a Polarized Congress, 1996–2015." *Environmental Politics* 30, no. 4 (2021): 538–58. https://doi/10.1080/09644016.2020.1786333.

Gunther, Marc. "Edelman Loses Executives and Clients Over Climate Change Stance." *The Guardian*, July 7, 2015. https://www.theguardian.com/sustainable-business/2015/jul/07/pr-edelman-climate-change-lost-executives-clients.

Gutiérrez, Miren. *Data Activism and Social Change*. London: Palgrave Pivot, 2018.

Habermas, Jürgen. "Political Communication in Media Society: Does Democracy Still Enjoy an Epistemic Dimension? The Impact of Normative Theory on Empirical Research." *Communication Theory* 16, no. 4 (2006): 411–26.

———. *The Structural Transformation of the Public Sphere: An Inquiry Into a Category of Bourgeois Society*. Trans. Thomas Burger. Cambridge, MA: MIT Press, 1991.

Hallin, Dan. "The Passing of the 'High Modernism' of American Journalism Revisited." *Political Communication Report* 16, no. 1 (2006): 14–25.

Handley, Steve. "Bill McKibben Calls FBI Tracking of Environmental Activists 'Contemptible.'" CleanTechnica, December 13, 2018. https://cleantechnica.com/2018/12/13/bill-mckibben-calls-fbi-tracking-of-environmental-activists-contemptible/.

Hao, Karen. "We Read the Paper That Forced Timnit Gebru out of Google. Here's What It Says." *MIT Technology Review*, December 4, 2020. https://www.technologyreview.com/2020/12/04/1013294/google-ai-ethics-research-paper-forced-out-timnit-gebru/.

Haraway, Donna. "A Cyborg Manifesto: Science, Technology, and Socialist-Feminism in the Late Twentieth Century." In *Simians, Cyborgs and Women: The Reinvention of Nature*, 117–58. New York: Routledge, 1991.

Harold, Christine. *Things Worth Keeping: The Value of Attachment in a Disposable World*. Minneapolis: University of Minnesota Press, 2020.

Hartley, John. *Understanding the News*. London: Methuen, 1982.

Harvey, Chelsea. "New IPCC Report Looks at Neglected Element of Climate Action: People." "E&E News," *Scientific American*, April 7, 2022. https://www.scientificamerican.com/article/new-ipcc-report-looks-at-neglected-element-of-climate-action-people/.

Haugen, Frances. "Opening Statement to Senate Committee on Commerce, Science & Transportation." *Frances Haugen* (blog), October 5, 2021. https://www.franceshaugen.com/blog/b9xlswihkike7639nn4ie230dz9eqy.

Hayes, Chris. "The New Abolitionism," *The Nation*, April 22, 2014. https://www.thenation.com/article/archive/new-abolitionism/.

Hayhoe, Katharine. *Saving Us: A Climate Scientist's Case for Hope and Healing in a Divided World*. New York: Simon and Schuster, 2021.

Hedrick, Ashley, Dave Karpf, and Daniel Kreiss. "The Earnest Internet vs. the Ambivalent Internet." *International Journal of Communication* 12 (2018): 1057–64.

Henn, Jamie. "Our Media Partnership with the *Guardian*." 350.org, April 1, 2015. https://350.org/our-media-partnership-with-the-guardian.

Hersher, Rebecca. "Key Moments in the Dakota Access Pipeline Fight." NPR, February 22, 2017. https://www.npr.org/sections/thetwo-way/2017/02/22/514988040/key-moments-in-the-dakota-access-pipeline-fight.

Hertsgaard, and Kyle Pope. "The Media Are Complacent While the World Burns." *Columbia Journalism Review*, April 22, 2019. https://www.cjr.org/special_report/climate-change-media.php.

Hiller, Jennifer, and Svea Herbst-Bayliss. "Exxon Loses Board Seats to Activist Hedge Fund in Landmark Climate Vote." Reuters, May 26, 2021. https://www.reuters.com/business/sustainable-business/shareholder-activism-reaches-milestone-exxon-board-vote-nears-end-2021-05-26/.

Hindman, Matthew. *The Internet Trap: How the Digital Economy Builds Monopolies and Undermines Democracy.* Princeton, NJ: Princeton University Press, 2018.

Hindman, Matthew, and Vladimir Barash. "Disinformation, 'Fake News' and Influence Campaigns on Twitter." Knight Foundation, October 4, 2018. https://knightfoundation.org/reports/disinformation-fake-news-and-influence-campaigns-on-twitter/.

Hintz, Arne, Lina Dencik, and Karin Wahl-Jorgensen. *Digital Citizenship in a Datafied Society.* Cambridge: Polity Press, 2018.

——. "Introduction." In "Digital Citizenship and Surveillance," special issue of *International Journal of Communication* 11 (2017): 731–39. https://doi.org/1932-8036/20170005.

Hirji, Zahra. "It's Not Just Greta. Trolls Are Swarming Young Climate Activists Online." *BuzzFeed News*, September 25, 2019. https://www.buzzfeednews.com/article/zahrahirji/greta-thunberg-climate-teen-activist-harassment.

Hirschfield, Paul J., and Simon Danella. "Legitimating Police Violence: Newspaper Narratives of Deadly Force." *Theoretical Criminology* 14, no. 2 (2010): 155–82. https://doi.org/10.1177/1362480609351545.

Hoppenstedt, Max. "Luisa Neubauer über Hatespeech: 'Was über Mich Geschrieben Wird, Ist Schon Krass.'" *Der Spiegel*, August 12, 2020. https://www.spiegel.de/netzwelt/netzpolitik/luisa-neubauer-ueber-hatespeech-was-ueber-mich-geschrieben-wird-ist-schon-krass-a-19bd80bb-4fec-4fea-8e11-85b817c05bf3.

Hulme, Mike. "Mediated Messages About Climate Change: Reporting the IPCC Fourth Assessment in the UK Print Media." *Climate Change and the Media* 2009:117–28.

Indianz. "Republican Investigation Links Russian Trolls to #NoDAPL Movement." March 1, 2018. https://www.indianz.com/News/2018/03/01/republican-investigation-links-russian-t.asp.

InfluenceMap. "Big Oil's Real Agenda on Climate Change." N.d. https://influencemap.org/report/How-Big-Oil-Continues-to-Oppose-the-Paris-Agreement-38212275958aa21196dae3b76220bddc.

———. "Big Tech and Climate Policy." January 2021. https://influencemap.org/report/Big-Tech-and-Climate-Policy-afb476c56f217ea0ab351d79096df04a.

———. "Climate Change and Digital Advertising: Climate Science Disinformation in Facebook Advertising." October 2020. https://influencemap.org/report/Climate-Change-and-Digital-Advertising-86222daed29c6f49ab2da76b0df15f76#4.

Integovernmental Panel on Climate Change. "Climate Change 2022: Impacts, Adaptation, and Vulnerability." N.d. https://www.ipcc.ch/report/ar6/wg2/.

Irwin, Luke. "The GDPR: Understanding the 6 Data Protection Principles." *IT Governance European Blog*, December 9, 2021. https://www.itgovernance.eu/blog/en/the-gdpr-understanding-the-6-data-protection-principles.

Ito, Mizuko. Introduction to *Networked Publics*, ed. Kazys Varnelis, 1–14. Cambridge, MA: MIT Press, 2008.

Jackson, Lauren. "Jack Dorsey on Twitter's Mistakes." *New York Times*, August 7, 2020. https://www.nytimes.com/2020/08/07/podcasts/the-daily/Jack-dorsey-twitter-trump.html.

Jamieson, Kathleen Hall, Dan Kahan, and Dietram A. Scheufele, eds. *The Oxford Handbook of the Science of Science Communication*. Oxford: Oxford University Press, 2017.

Jefferson, Whitney. "Here's How Celebrities Are Taking Part in the Climate Strike." *BuzzFeed*, November 12, 2021. https://www.buzzfeed.com/whitneyjefferson/celebrities-taking-part-in-the-climate-strike.

Jenkins, Henry. *Convergence Culture: Where Old and New Media Collide*. Cambridge, MA: MIT Press, 2006.

Jenkins, Joy, and Lucas Graves. *Case Studies in Collaborative Local Journalism*. Reuters Institute Report. Oxford: Reuters Institute for the Study of Journalism, April 25, 2019. https://reutersinstitute.politics.ox.ac.uk/our-research/case-studies-collaborative-local-journalism.

Joselow, Maxine, and Dino Grandoni. "Big Oil CEOs Testify Before House Oversight Committee." *Washington Post*, October 28, 2021. https://www

.washingtonpost.com/climate-environment/2021/10/28/oil-executives-testimony-live-updates/.

Kahn, Brian. "Is This a Joke?" *Gizmodo*, September 15, 2020. https://earther.gizmodo.com/is-this-a-joke-1845063054.

Kaufman, Mark. "The Carbon Footprint Scam." *Mashable*, August 2021. https://mashable.com/feature/carbon-footprint-pr-campaign-sham.

Kingaby, Harriet. "Climate Disinformation: A Beginner's Guide." *Branch*, no. 2 (Spring 2021). https://branch.climateaction.tech/issues/issue-2/climate-disinformation-a-beginners-guide/.

Knaus, Christopher. "Bots and Trolls Spread False Arson Claims in Australian Fires 'Disinformation Campaign.'" *The Guardian*, January 7, 2020. https://www.theguardian.com/australia-news/2020/jan/08/twitter-bots-trolls-australian-bushfires-social-media-disinformation-campaign-false-claims.

Koebler, Jason. "The Government Is Using a No Fly Zone to Suppress Journalism at Standing Rock." *Vice*, November 30, 2016. https://www.vice.com/en/article/yp3kak/the-government-is-using-a-no-fly-zone-to-suppress-journalism-at-standing-rock.

Krange, Olve, Bjørn P. Kaltenborn, and Martin Hultman. "'Don't Confuse Me with Facts'—How Right Wing Populism Affects Trust in Agencies Advocating Anthropogenic Climate Change as a Reality." *Humanities and Social Sciences Communications* 8, no. 1 (2021): 1–9. https://www.nature.com/articles/s41599-021-00930-7.

Kreiss, Daniel. *Prototype Politics: Technology-Intensive Campaigning and the Data of Democracy*. Oxford: Oxford University Press, 2016.

Kunelius, Risto. "A Forced Opportunity: Climate Change and Journalism." *Journalism* 20, no. 1 (2019): 218–21. https://doi.org/10.1177/1464884918807596.

Kunelius, Risto, and Elisabeth Eide. "Moment of Hope, Mode of Realism: On the Dynamics of a Transnational Journalistic Field During UN Climate Change Summits." *International Journal of Communication* 6 (2012): 266–85. https://link.gale.com/apps/doc/A287109746/AONE?u=wash_main&sid=AONE&xid=4d20ad0b.

Kunelius, Risto, Elisabeth Eide, Matthew Tegelberg, and Dmitry Yagodin, eds. *Media and Global Climate Knowledge: Journalism and the IPCC*. New York: Palgrave MacMillan, 2017.

Lakhani, Nina. "U.S. Supreme Court Rejects Dakota Access Pipeline Appeal." *The Guardian,* February 22, 2022. https://www.theguardian.com/us-news/2022/feb/22/us-supreme-court-dakota-access-pipeline.

Latour, Bruno. *Reassembling the Social: An Introduction to Actor-Network-Theory.* Oxford: Oxford University Press, 2005.

Lavelle, Marianne. "'Trollbots' Swarm Twitter with Attacks on Climate Science Ahead of UN Summit." *Inside Climate News,* September 16, 2019. https://insideclimatenews.org/news/16092019/trollbot-twitter-climate-change-attacks-disinformation-campaign-mann-mckenna-greta-targeted/.

Leber, Rebecca. "What the Oil Industry Still Won't Tell Us." *Vox,* October 28, 2021. https://www.vox.com/22745597/big-oil-congress-hearing-exxonmobil-bp-chevron-shell.

Lee, Ashley. "Invisible Networked Publics and Hidden Contention: Youth Activism and Social Media Tactics Under Repression." *New Media & Society* 20, no. 11 (November 2018): 4095–115. https://doi.org/10.1177/1461444818768063.

Leigh, Andrew. "How Populism Imperils the Planet." *MIT Press Reader,* November 5, 2021. https://thereader.mitpress.mit.edu/how-populism-imperils-the-planet/.

Leiserowitz, Anthony, Edward Maibach, Seth Rosenthal, John Kotcher, Jennifer Carman, Xinran Wang, Matthew Goldberg, et al. *Climate Change in the American Mind: December 2020.* New Haven, CT: Yale Program on Climate Change Communication, February 10, 2021. https://climatecommunication.yale.edu/publications/climate-change-in-the-american-mind-december-2020/.

Levine, Alexandra S., Claire Rafford, Julia Arciga, Emily Birnbaum, and Benjamin Din. "Whistleblower to Senate: Don't Trust Facebook." *Politico,* October 5, 2021. https://www.politico.com/news/2021/10/05/facebook-whistleblower-testifies-congress-515083.

Lewis, Seth C., and Logan Molyneux. "A Decade of Research on Social Media and Journalism: Assumptions, Blind Spots, and a Way Forward." *Media and Communication* 6, no. 4 (2018): 11–23.

Lewis, Seth C., Rodrigo Zamith, and Mark Coddington. "Online Harassment and Its Implications for the Journalist–Audience Relationship." *Digital Journalism* 8, no. 8 (2020): 1047–67. https://doi.org/10.1080/21670811.2020.1811743.

Lievrouw, Leah. "Materiality and Media in Communication and Technology Studies: An Unfinished Project." In *Media Technologies: Essays on Communication, Materiality, and Society*, ed. Tarleton Gillespie, Pablo J. Boczkowski, and Kirstin A. Foot, 21–51. Cambridge, MA: MIT Press 2014.

Lifton, Robert Jay. *The Climate Swerve: Reflections on Mind, Hope, and Survival.* New York: New Press, 2017.

Lim, Audrea. "The Ideology of Fossil Fuels." *Dissent*, Spring 2018. https://www.dissentmagazine.org/article/ideology-fossil-fuels-apocalypse-petrocapitalism-energy-humanities.

Lippmann, Walter. *The Phantom Public: A Sequel to "Public Opinion."* New York: Macmillan, 1927.

———. *Public Opinion.* New York: Harcourt, Brace, 1922.

Lück, Julia, Antal Wozniak, and Hartmut Wessler. "Networks of Coproduction: How Journalists and Environmental NGOs Create Common Interpretations of the UN Climate Change Conferences." *International Journal of Press/Politics* 21, no. 1 (2016): 25–47. https://doi.org/10.1177/1940161215612204.

Lush. "Stop the Pipelines—No More Tar Sand!" N.d. https://www.lushusa.com/stories/article_ethical-stop-the-pipelines.html.

MacKendrick, Norah. "Out of the Labs and Into the Streets: Scientists Get Political." *Sociological Forum* 32, no. 4 (2017): 896–902. https://doi.org/10.1111/socf.12366.

Manoff, Robert Karl, and Michael Schudson, eds. *Reading the News: A Pantheon Guide to Popular Culture.* New York: Pantheon, 1986.

Marchi, Regina, and Lynn Schofield Clark. "Social Media and Connective Journalism: The Formation of Counterpublics and Youth Civic Participation." *Journalism* 22, no. 2 (2021): 285–302.

Marres, Noortje. "Issues Spark a Public Into Being: A Key but Often Forgotten Point of the Lippman–Dewey Debate." In *Making Things Public: Atmospheres of Democracy*, ed. Bruno Latour and Peter Weibel, 208–17. Cambridge, MA: MIT Press, 2005.

———. "Why We Can't Have Our Facts Back." *Engaging Science, Technology, and Society* 4 (2018): 423–43. https://doi.org/10.17351/ests2018.188.

Marwick, Alice, and Rebecca Lewis. "Media Manipulation and Disinformation Online." *Data & Society*, May 15, 2017. https://datasociety.net/library/media-manipulation-anddisinfo-online.

Matthews, Christopher M. "Silicon Valley to Big Oil: We Can Manage Your Data Better Than You." *Wall Street Journal*, July 24, 2018. https://www.wsj.com/articles/silicon-valley-courts-a-wary-oil-patch-1532424600.

Mazaska Talks. "Divest Your Community." N.d. https://mazaskatalks.org/divest-your-community.

McAllister, Lucy, Meaghan Daly, Patrick Chandler, Marisa McNatt, Andrew Benham, and Maxwell Boykoff. "Balance as Bias, Resolute on the Retreat? Updates and Analyses of Newspaper Coverage in the United States, United Kingdom, New Zealand, Australia and Canada Over the Past 15 Years." *Environmental Research Letters* 16, no. 9 (2021): 1–14. https://doi.org/10.1088/1748-9326/ac14eb.

McCright, Aaron M., and Riley E. Dunlap. "The Politicization of Climate Change and Polarization in the American Public's Views of Global Warming, 2001–2010." *Sociological Quarterly* 52, no. 2 (2011): 155–94. https://doi.org/10.1111/j.1533-8525.2011.01198.

McFall-Johnsen, Morgan. "The Companies Polluting the Planet Have Spent Millions to Make You Think Carpooling and Recycling Will Save Us." *Business Insider*, September 18, 2021. https://www.businessinsider.com/fossil-fuel-companies-spend-millions-to-promote-individual-responsibility-2021-3.

McKibben, Bill. "CJR's Covering Climate Change." YouTube video, May 17, 2019, 5:18:43. https://www.youtube.com/watch?v=FO9DKk07SCY.

——. "This Movement Is Taking Money Away from Fossil Fuels, and It's Working." *New York Times*, October 26, 2021. https://www.nytimes.com/2021/10/26/opinion/climate-change-divestment-fossil-fuels.html.

Media and Climate Change Observatory. "The Sobering Realization That We're Going Completely in the Wrong Direction." 58 (October 2021). https://sciencepolicy.colorado.edu/icecaps/research/media_coverage/summaries/issue58.html.

Merchant, Brian. "How Google, Microsoft, and Big Tech Are Automating the Climate Crisis." *Gizmodo*, February 21, 2019. https://gizmodo.com/how-google-microsoft-and-big-tech-are-automating-the-1832790799.

Milan, Stefania. "The Materiality of Clouds: Beyond a Platform-Specific Critique of Contemporary Activism." In *Social Media Materialities and Protest: Critical Reflections*, ed. Mette Mortensen, Christina Neumayer, and Thomas Poell, 116–27. London: Routledge, 2019.

Monet, Jenni. "Covering Standing Rock." *Columbia Journalism Review*, Spring 2017. https://www.cjr.org/local_news/covering-standing-rock.php.

Moore, Martin, and Damian Tambini. *Regulating Big Tech: Policy Responses to Digital Dominance*. Oxford: Oxford University Press, 2021.

Mortensen, Mette, Christina Neumayer, and Thomas Poell, eds. *Social Media Materialities and Protest: Critical Reflections* London: Routledge, 2019.

Myers, Teresa, Matthew Nisbet, Edward Maibach, and Anthony Leiserowitz. "A Public Health Frame Arouses Hopeful Emotions About Climate Change." *Climatic Change* 113, nos. 3–4 (2012): 1105–12. https://doi.org.10.1007/s10584-012-0513-6.

Nadim, Marjan, and Audun Fladmoe. "Silencing Women? Gender and Online Harassment." *Social Science Computer Review* 39, no. 2 (2021): 245–58. https://doi.org/10.1177/0894439319865518.

Nadler, Anthony. "Nature's Economy and News Ecology." *Journalism Studies* 20, no. 6 (2019): 823–39. https://doi.org/10.1080/1461670X.2018.1427000.

Nelson, Jacob. "A Twitter Tightrope Without a Net: Journalists' Reactions to Newsroom Social Media Policies." *Columbia Journalism Review*, December 2, 2021. https://www.cjr.org/tow_center_reports/newsroom-social-media-policies.php.

Nelson, Milla, and Meg Vertigan. "Misogyny, Male Rage and the Words Men Use to Describe Greta Thunberg." *The Conversation*, November 12, 2021. http://theconversation.com/misogyny-male-rage-and-the-words-men-use-to-describe-greta-thunberg-124347.

Neumeyer, Christina, Mette Mortensen, and Thomas Poell. "Introduction: Social Media Materialities and Protest." In *Social Media Materialities and Protest: Critical Reflections*, ed. Mette Mortensen, Christina Neumayer, and Thomas Poell, 1–14. London: Routledge, 2019.

Newman, Nic, with Richard Fletcher, Anne Schulz, Simge Andı, and Rasmus Kleis Nielsen. *Digital News Report 2020*. Oxford: Reuters Institute for the Study of Journalism, 2020. https://reutersinstitute.politics.ox.ac.uk/sites/default/files/2020-06/DNR_2020_FINAL.pdf.

Newman, Todd P., Erik C. Nisbet, and Matthew C. Nisbet. "Climate Change, Cultural Cognition, and Media Effects: Worldviews Drive News Selectivity, Biased Processing, and Polarized Attitudes." *Public

Understanding of Science 27, no. 8 (2018): 985–1002. https://doi.org/10.1177/0963662518801170.

Newton, Casey, "How Facebook Rewards Polarizing Political Ads." *The Verge*, October 11, 2017. https://www.theverge.com/2017/10/11/16449976/facebook-political-ads-trump-russia-election-news-feed.

Nicanor, Nepeti. "Vanessa Nakate's Erasure Portrays an Idealised Climate Activism." *Africa at LSE* (blog), January 31, 2020. https://blogs.lse.ac.uk/africaatlse/2020/01/31/vanessa-nakate-davos-cropped-photo-white-race-climate-activism/.

Nielsen, Rasmus Kleis, and Sarah Anne Ganter. *The Power of Platforms: Shaping Media and Society*. Oxford: Oxford University Press, 2022.

Nielsen, Rasmus Kleis, and Meera Selva. *More Important, but Less Robust? Five Things Everybody Needs to Know About the Future of Journalism*. Reuters Institute Report. Oxford: Reuters Institute for the Study of Journalism, January 2019. https://reutersinstitute.politics.ox.ac.uk/our-research/more-important-less-robust-five-things-everybody-needs-know-about-future-journalism.

Nisbet, Matthew C. "Communicating Climate Change: Why Frames Matter for Public Engagement." *Environment: Science and Policy for Sustainable Development* 41, no. 2 (2009): 12– 23. https://doi.org/10.3200/ENVT.51.2.12-23.

———. "Nature's Prophet: Bill McKibben as Journalist, Public Intellectual and Activist." Shornstein Center on Media, Politics, and Public Policy, March 7, 2013. shorensteincenter.org/wp-content/uploads/2013/03/D-78-Nisbet1.pdf.

Nishime, Leilani, and Kim D. Hester Williams, eds. *Racial Ecologies*. Seattle: University of Washington Press, 2018.

Noble, Safiya. *Algorithms of Oppression: How Search Engines Reinforce Racism*. New York: New York University Press, 2018.

Nosek, Grace. "The Fossil Fuel Industry's Push to Target Climate Protesters in the U.S." *Pace Environmental Law Review* 38 (2020), https://digitalcommons.pace.edu/pelr/vol38/iss1/2.

Oreskes, Naomi. "Beyond the Ivory Tower: The Scientific Consensus on Climate Change." *Science* 306, no. 5702 (2004): 1686. https://doi.org/10.1126/science.1103618.

Oreskes, Naomi, and Erik M. Conway. *The Collapse of Western Civilization: A View from the Future*. New York: Columbia University Press, 2014.

———. *Merchants of Doubt: How a Handful of Scientists Obscured the Truth on Issues from Tobacco Smoke to Climate Change.* New York: Bloomsbury, 2011.

Outcalt, Chris, and Brittany Peterson. "Colorado River 100 Years." *AP News*, September 12, 2022. https://apnews.com/hub/colorado-river-100-years.

Overpass Light Brigade (blog). "Drone to Be Wild." N.d. http://overpass lightbrigade.org/drone-to-be-wild/.

Painter, James, María Carmen Erviti, Richard Fletcher, Candice Howarth, Silje Kristiansen, Leon Bienvenido, Alan Oukrat, et al. *Something Old, Something New: Digital Media and the Coverage of Climate Change.* Oxford: Reuters Institute for the Study of Journalism, 2016.

Palmieri, Edward. "The Next Decade: How Facebook Is Stepping Up the Fight Against Climate Change." *Facebook Engineering*, September 14, 2020. https://engineering.fb.com/2020/09/14/data-center-engineering/net-zero-carbon/.

Papacharissi, Zizi. *Affective Publics: Sentiment, Technology and Politics.* Oxford: Oxford University Press, 2015.

———. *After Democracy: Imagining Our Political Future.* New Haven, CT: Yale University Press, 2021.

Pariser, Eli. *The Filter Bubble: How the New Personalized Web Is Changing What We Read and How We Think.* New York: Penguin, 2011.

Pasquale, Frank A. "The Automated Public Sphere." University of Maryland Legal Studies Research Paper, no. 2017-31, November 8, 2017. https://ssrn.com/abstract=3067552.

———. "Internet Nondiscrimination Principles Revisited." *SSRN Electronic Journal*, no. 629 (2020): 1–39. https://doi.org/10.2139/ssrn.3634625.

Pattee, Emily. "Meet the Climate Change Activists of TikTok." *Wired*, March 11, 2021. https://www.wired.com/story/climate-change-tiktok-science-communication/.

Peters, Adele. "Inside the Campaign to Push PR Firms to Fire Fossil Fuel Clients." *Fast Company*, November 25, 2020. https://www.fastcompany.com/90579261/inside-the-campaign-to-push-pr-firms-to-fire-fossil-fuel-clients.

Pezzullo, Phaedra C., and Robert Cox. *Environmental Communication and the Public Sphere.* 5th ed. Thousand Oaks, CA: SAGE, 2018.

Phillips, Whitney. "Navigating the Information Landscape: A Media Literacy Toolkit Series." *Commonplace*, June 30, 2020. https://doi.org/10.21428/6ffd0432.3/155cc0.

———. *The Oxygen of Amplification: Better Practices for Reporting on Extremists, Antagonists, and Manipulators Online*. New York: Data & Society Research Institute, 2018. https://datasociety.net/wp-content/uploads/2018/05/0-EXEC-SUMMARY_Oxygen_of_Amplification_DS-1.pdf.

Phillips, Whitney, and Ryan M. Milner. *The Ambivalent Internet: Mischief, Oddity, and Antagonism Online*. London: Wiley, 2018.

———. *You Are Here: A Field Guide for Navigating Polarized Speech, Conspiracy Theories, and Our Polluted Media Landscape*. Cambridge, MA: MIT Press, 2021.

Pickard, Victor. *America's Battle for Democracy: The Triumph of Corporate Libertarianism and the Future of Media Reform*. New York: Cambridge University Press, 2015.

———. "The Big Picture: Misinformation Society." *Public Books*, November 28, 2017. https://www.publicbooks.org/the-big-picture-misinformation-society/.

———. *Democracy Without Journalism? Confronting the Misinformation Society*. Oxford: Oxford University Press, 2019.

———. "Restructuring Democratic Infrastructures: A Policy Approach to the Journalism Crisis." *Digital Journalism* 8, no. 6 (2020): 704–19. https://doi.org/10.1080/21670811.2020.1733433.

———. "The Violence of the Market." *Journalism* 20, no. 1 (2019): 154–58. https://doi.org/10.1177/1464884918808955.

Pinto, Juliet, Robert E. Gutsche, and Paola Prado, eds. *Climate Change, Media & Culture: Critical Issues in Global Environmental Communication*. Bradford, U.K.: Emerald, 2019.

Plantin, Jean-Christophe, and Aswin Punathambekar. "Digital Media Infrastructures: Pipes, Platforms, and Politics." *Media, Culture & Society* 41, no. 2 (2019): 163–74.

Pope, Kyle. "Giving Climate the Coverage It Deserves." *Columbia Journalism Review*, June 15, 2022. https://www.cjr.org/covering_climate_now/giving-climate-the-coverage-it-deserves.php.

Powers, Matthew, and Adrienne Russell, eds. *Rethinking Media Research for Changing Societies*. Cambridge: Cambridge University Press, 2020.

Ramirez, Rachel. "Spanning Beats, Environmental Justice Reporting Influences Every Story." *NiemanReports*, February 3, 2021. https://niemanreports

.org/articles/spanning-beats-environmental-justice-reporting-influences-every-story/.

Reed, Stanley, and Claire Moses. "A Dutch Court Rules That Shell Must Step Up Its Climate Change Efforts." *New York Times*, October 28, 2021. https://www.nytimes.com/2021/05/26/business/royal-dutch-shell-climate-change.html.

Reyes, Maria, and Adriana Calderón. "What Is MAPA and Why Should We Pay Attention to It?" Fridays for Future blog, March 13, 2021. https://fridaysforfuture.org/newsletter/edition-no-1-what-is-mapa-and-why-should-we-pay-attention-to-it/.

Rez, Guy. "How I Built Resilience: Varshini Prakash of Sunrise Movement." *How I Built This* (podcast), NPR, November 5, 2020. https://www.npr.org/2020/10/28/928810660/how-i-built-resilience-varshini-prakash-of-sunrise-movement.

Richardson, Allissa V. *Bearing Witness While Black: African Americans, Smartphones, and the New Protest #Journalism*. New York: Oxford University Press, 2020.

Rieder, Bernhard, Ariadna Matamoros-Fernández, and Òscar Coromina. "From Ranking Algorithms to 'Ranking Cultures': Investigating the Modulation of Visibility in YouTube Search Results." *Convergence* 24, no. 1 (2018): 50–68. https://doi.org/10.1177/1354856517736982.

Ripple, William, et al. "World Scientists' Warning of a Climate Emergency." *BioScience* 70, no. 1 (January 2020): 8–12.

Roosvall, Anna, and Matthew Tegelberg. *Media and Transnational Climate Justice Indigenous Activism and Climate Politics*. New York: Peter Lang, 2017.

Russell, Adrienne. "Coming to Terms with Dysfunctional Hybridity: A Conversation with Andrew Chadwick on the Challenges to Liberal Democracy in the Second-Wave Networked Era." *Studies in Communication Sciences* 20, no. 2 (2020): 211–25.

——. "Innovation in Hybrid Spaces: 2011 UN Climate Summit and the Changing Journalism Field." *Journalism* 14, no. 7 (2013): 904–20.

——. *Journalism as Activism: Recoding Media Power*. Cambridge: Polity, 2016.

Russell, Adrienne, Mizuki Ito, Todd Richmond, and Marc Tuters. "Culture: Media Convergence and Networked Participation." In *Networked Publics*, ed. Kazys Varnelis, 43–76. Cambridge, MA: MIT Press, 2008.

Russell, Adrienne, Jarkko Kangas, Risto Kunelius, and James Painter. "Niche Climate News Sites and the Changing Context of Covering Catastrophe." *Journalism: Theory, Practice, Criticism*, online, July 6, 2022. https://doi.org/10.1177/14648849221113119.

Russell, Adrienne, and Matthew Tegelberg. "Beyond the Boundaries of Science: Resistance to Misinformation by Citizen Scientists." *Journalism: Theory, Practice, Criticism* 21, no. 3 (2019): 327–44. https://doi.org/10.1177/1464884919862655.

Salmón, Enrique. "Kincentric Ecology: Indigenous Perceptions of the Human–Nature Relationship." *Ecological Applications* 10, no. 5 (2000): 1327–32. https://doi.org/10.1890/1051-0761(2000)010[1327:KEIPOT]2.0.CO;2.

Sanson, Ann V., Judith Van Hoorn, and Susie E. L. Burke. "Responding to the Impacts of the Climate Crisis on Children and Youth." *Child Development Perspectives* 13, no. 4 (2019): 201–7. https://doi.org/10.1111/cdep.12342.

Schäfer, Mike S., and Inga Schlichting. "Media Representations of Climate Change: A Meta-analysis of the Research Field." *Environmental Communication* 8, no. 2 (2014): 142–60. https://doi.org/10.1080/17524032.2014.914050.

Schilling, Vincent. "No DAPL Social Media Explosion: Celebrities, Musicians and Politicians Worldwide." *Indian Country Today*, September 11, 2016. https://indiancountrytoday.com/archive/nodapl-social-media-explosion-celebrities-musicians-and-politicians-worldwide.

Schudson, Michael. *Discovering the News: A Social History of American Newspapers*. New York: Basic, 1978.

———. *The Sociology of News*. New York: Norton, 2003.

Schwartz, John. "Katharine Hayhoe, a Climate Explainer Who Stays Above the Storm." *New York Times*, October 10, 2016. https://www.nytimes.com/2016/10/11/science/katharine-hayhoe-climate-change-science.html.

Schwartzstein, Peter. "The Authoritarian War on Environmental Journalism." Century Foundation, July 7, 2020. https://tcf.org/content/report/authoritarian-war-environmental-journalism/.

Scott, Ellen. "Meet the Teens Making Climate Change Memes to Deal with Ecoanxiety." *Metro* (blog), August 15, 2019. https://metro.co.uk/2019/08/15/meet-teens-making-climate-change-memes-deal-ecoanxiety-10570574/.

Shoemaker, Pamela J. "Media Treatment of Deviant Political Groups." *Journalism Quarterly* 61, no. 1 (1984): 66–82.

Sinclair, Upton. *The Jungle: The Uncensored Original Edition*. Ed. Earl Lee. Tucson, AZ: See Sharp Press, 2003.

Smith, Adam B. "2020 U.S. Billion-Dollar Weather and Climate Disasters in Historical Context." Climate.gov, 2021. https://www.climate.gov/disasters2020.

Squires, Catherine R. "Rethinking the Black Public Sphere: An Alternative Vocabulary for Multiple Public Spheres." *Communication Theory* 12, no. 4 (2002): 446–68.

Stand.earth. "1485 Institutions with Assets Over $39.2 Trillion Have Committed to Divest from Fossil Fuels." October 26, 2021. https://www.stand.earth/divestinvest2021.

Stanley-Becker, Isaac. "Trump, Pressed on the Environment in U.K. Visit, Says Climate Change Goes 'Both Ways.'" *Washington Post*, June 5, 2019. https://www.washingtonpost.com/world/europe/trump-pressed-on-the-environment-in-uk-visit-says-climate-change-goes-both-ways/2019/06/05/77c8750c-8717-11e9-9d73-e2ba6bbf1b9b_story.html.

Stark, Birgit, Daniel Stegmann, Melanie Magin, and Pascal Jürgens. "Are Algorithms a Threat to Democracy? The Rise of Intermediaries: A Challenge for Public Discourse." Algorithm Watch, May 2020. https://algorithmwatch.org/en/wp-content/uploads/2020/05/Governing-Platforms-communications-study-Stark-May-2020-AlgorithmWatch.pdf.

Stoll, Mark. *Rachel Carson's* Silent Spring, *a Book That Changed the World. Environment and Society*, virtual exhibition, 2012. http://www.environmentandsociety.org/exhibitions/silent-spring/overview.

Sunstein, Cass R. *#Republic: Divided Democracy in the Age of Social Media*. Princeton, NJ: Princeton University Press, 2018.

Supran, Geoffery, and Naomi Oreskes. "The Forgotten Oil Ads That Told Us Climate Change Was Nothing." *The Guardian*, November 18, 2021. https://www.theguardian.com/environment/2021/nov/18/the-forgotten-oil-ads-that-told-us-climate-change-was-nothing.

Szalai, Jennifer. "In 'The Uninhabitable Earth,' Apocalypse Is Now." *New York Times*, March 6, 2019. https://www.nytimes.com/2019/03/06/books/review-uninhabitable-earth-life-after-warming-david-wallace-wells.html.

Tabuchi, Hiroko, and Nadja Popovich. "People of Color Breathe More Hazardous Air. The Sources Are Everywhere." *New York Times*, April 28, 2021. https://www.nytimes.com/2021/04/28/climate/air-pollution-minorities.html.

Taft, Molly. "Chevron Jumps Into Texas' News Desert with Stories About Puppies, Football, and Oil." *Gizmodo*, August 18, 2022. https://gizmodo.com/chevron-local-news-texas-permian-proud-1849424317.

——. "For Earth Day, Houston Public Media Is Promoting . . . Chevron?" *Gizmodo*, April 21, 2022. https://gizmodo.com/houston-public-media-chevron-partnership-earth-day-1848822428.

Tambini, Damian. *Media Freedom*. Cambridge: Polity, 2021.

Tandoc, Edson C., and Bruno Takahashi. "Playing a Crusader Role or Just Playing by the Rules? Role Conceptions and Role Inconsistencies Among Environmental Journalists." *Journalism* 15, no. 7 (2014): 889–907. https://doi.org/10.1177/1464884913501836.

Taylor, Matthew. "Environment Protest Being Criminalised Around World, Say Experts." *The Guardian*, April 29, 2021. https://www.theguardian.com/environment/2021/apr/19/environment-protest-being-criminalised-around-world-say-experts.

Tech Workers Coalition. "Climate Strike." N.d. https://techworkerscoalition.org/climate-strike/.

——. "Worker Power in the Tech Industry." N.d. https://techworkerscoalition.org/.

Telford, Taylor. "These Self-Described Trolls Tackle Climate Disinformation on Social Media with Wit and Memes." *Washington Post*, July 30, 2021. https://www.washingtonpost.com/business/2021/07/30/greentrolling-big-oil-greenwashing/.

Tepper, Jonathan, and Denise Hearn. *The Myth of Capitalism: Monopolies and the Death of Competition*. Newark, NJ: Wiley, 2018.

Tessum, Christopher W., David A. Paoella, Sarah E. Chambliss, Joshua S. Apte, Jason D. Hill, and Julian D. Marshall. "PM2.5 Polluters Disproportionately and Systemically Affect People of Color in the United States." *Science Advances* 7, no. 18 (2021): 1–6. https://doi.org/10.1126/sciadv.abf4491.

Thompson, Clive. "We Might Be Reaching 'Peak Indifference' on Climate Change." *Wired*, March 25, 2019. https://www.wired.com/story/we-might-be-reaching-peak-indifference-on-climate-change/.

Thunberg, Greta, Adriana Calderón, Farzana Faruk Jhumu, and Eric Njuguna. "This Is the World Being Left to Us by Adults." *New York Times*, August 19, 2021. https://www.nytimes.com/2021/08/19/opinion/climate-un-report-greta-thunberg.html.

Timberg, Craig, and Tomy Romm. "Facebook CEO Mark Zuckerberg to Capitol Hill: 'It Was My Mistake, and I'm Sorry.'" *Washington Post*, April 9, 2018. https://www.washingtonpost.com/news/the-switch/wp/2018/04/09/facebook-chief-executive-mark-zuckerberg-to-captiol-hill-it-was-my-mistake-and-im-sorry/.

———. "These Provocative Images Show Russian Trolls Sought to Inflame Debate Over Climate Change, Fracking and Dakota Pipeline." *Washington Post*, March 1, 2018. https://www.washingtonpost.com/news/the-switch/wp/2018/03/01/congress-russians-trolls-sought-to-inflame-u-s-debate-on-climate-change-fracking-and-dakota-pipeline/.

Tracy, Marc. "McClatchy, Family-Run News Chain, Goes to Hedge Fund in Bankruptcy Sale." *New York Times*, August 4, 2020. https://www.nytimes.com/2020/08/04/business/media/mcclatchy-newspapers-bankrutpcy-chatham.html.

Treisman, Rachel. "Facebook Fell Short of Its Promises to Label Climate Change Denial, a Study Finds." NPR, February 23, 2022. https://www.npr.org/2022/02/23/1082561725/facebook-climate-change-label.

Treré, Emiliano. *Hybrid Media Activism: Ecologies, Imaginaries, Algorithms*. New York: Routledge, 2018.

Trottier, Daniel, and Christian Fuchs. "Theorising Social Media, Politics and the State." In *Social Media, Politics and the State: Protests, Revolutions, Riots, Crime and Policing in the Age of Facebook, Twitter, and YouTube*, ed. Daniel Trottier and Christian Fuchs, 15–50. New York: Routledge, 2015.

Truth Tobacco Industry Documents. "Smoking and Health Proposal." N.d. https://www.industrydocuments.ucsf.edu/tobacco/docs/#id=psdw0147.

Truvill, William. "The News 50: Tech Giants Dwarf Murdoch." *Press Gazette* (United Kingdom), December 3, 2020. https://pressgazette.co.uk/biggest-media-companies-world/.

Tufekci, Zeynep. "Facebook's Surveillance Machine." *New York Times*, March 19, 2018. https://www.nytimes.com/2018/03/19/opinion/facebook-cambridge-analytica.html.

———. *Twitter and Tear Gas: The Power and Fragility of Networked Protest*. New Haven, CT: Yale University Press, 2017.

———. "YouTube, the Great Radicalizer." *New York Times*, March 10, 2018. https://www.nytimes.com/2018/03/10/opinion/sunday/youtube-politics-radical.html.

Turner, Fred. *From Counterculture to Cyberculture: Stewart Brand, the Whole Earth Network, and the Rise of Digital Utopianism*. Chicago: University of Chicago Press, 2006.

———. "The World Outside and the Pictures in Our Networks." In *Media Technologies: Essays on Communication, Materiality, and Society*, ed. Tarleton Gillespie, Pablo J. Boczkowski, and Kirsten A. Foot, 251–60. Cambridge, MA: MIT Press, 2014.

Turow, Joseph. *The Daily You: How the New Advertising Industry Is Defining Your Identity and Your Worth*. New Haven, CT: Yale University Press, 2012.

Turow, Joseph, and Nick Couldry. "Media as Data Extraction: Towards a New Map of a Transformed Communications Field." *Journal of Communication* 68, no. 2 (2018): 415–23. https://doi.org/10.1093/joc/jqx011.

USC Annenberg, School for Communication and Journalism. "In 'Algorithms of Oppression,' Safiya Noble Finds Old Stereotypes Persist in New Media." February 16, 2018, updated December 11, 2019. https://annenberg.usc.edu/news/diversity-and-inclusion/algorithms-oppression-safiya-noble-finds-old-stereotypes-persist-new.

U.S. Senate, Committee on Commerce, Science, and Transportation. "At Hearing with Big Tech CEOs, Cantwell Defends Local Journalism, Presses Platforms on Unfair Practices." October 2010. https://www.commerce.senate.gov/2020/10/at-hearing-with-big-tech-ceos-cantwell-defends-local-journalism-presses-platforms-on-unfair-practices.

Valentino-DeVries, Jennifer, Jeremy Singer-Vine, and Ashkan Soltani. "Websites Vary Prices, Deals Based on Users' Information." *Wall Street Journal*, December 24, 2012. https://www.wsj.com/articles/SB10001424127887323777204578189391813881534.

Vargas, Jose Antonio. "The Face of Facebook." *The New Yorker*, September 13, 2010. https://www.newyorker.com/magazine/2010/09/20/the-face-of-facebook.

Varnelis, Kazys, ed. *Networked Publics*. Cambridge, MA: MIT Press, 2007.

Vermeulen, Mathias. "Regulating the Digital Public Sphere." Open Society Foundations, June 2021. https://www.opensocietyfoundations.org/uploads/26fac926-c4ce-415e-ae85-a09fcbe10ba4/regulating-the-digital-public-sphere-report-20210617.pdf.

Vincent, Emmanuel. "Scientists Explain What *New York Magazine* Article on 'The Uninhabitable Earth' Gets Wrong." *Climate Feedback*, July 12, 2017. https://climatefeedback.org/evaluation/scientists-explain-what-new-york-magazine-article-on-the-uninhabitable-earth-gets-wrong-david-wallace-wells.

Vo, Lam Thuy. "Breaking Free from the Tyranny of the Loudest." Nieman-Lab blog, December 9, 2017. https://www.niemanlab.org/2017/12/breaking-free-from-the-tyranny-of-the-loudest.

Wahl-Jorgensen, Karin. "Questioning the Ideal of the Public Sphere: The Emotional Turn." *Social Media + Society* 5, no. 3 (2019). https://doi.org/10.1177/2056305119852175.

Waisbord, Silvio. "Antipress Violence and the Crisis of the State." *Harvard International Journal of Press/Politics* 7, no. 3 (2002): 90–109. https://doi.org/10.1177/1081180X0200700306.

Wakabayashi, Daisuke, and Tiffany Hsu. "Why Google Backtracked on Its New Search Results Look." *New York Times*, January 31, 2020. https://www.nytimes.com/2020/01/31/technology/google-search-results.html.

Waldman, Scott. "Climate Denial Spreads on Facebook as Scientists Face Restrictions." *ClimateWire*, July 6, 2020. https://www.scientificamerican.com/article/climate-denial-spreads-on-facebook-as-scientists-face-restrictions/.

———. "How CO_2 Boosters' Op-Ed Slipped by Facebook Fact-Checkers." *ClimateWire*, June 23, 2020. https://www.eenews.net/stories/1063436369.

Waldman, Scott, and Niina Heikkinen. "As Climate Scientists Speak Out, Sexist Attacks Are on the Rise." "E&E News," *Scientific American*, August 22, 2018. https://www.scientificamerican.com/article/as-climate-scientists-speak-out-sexist-attacks-are-on-the-rise/.

Walinchus, Lucia. "We Need a News Utility." *Poynter* (blog), June 16, 2022. https://www.poynter.org/commentary/2022/national-tax-support-journalism/.

Wallace-Wells, David. "The Uninhabitable Earth." *New York Magazine*, July 9, 2017. https://nymag.com/intelligencer/2017/07/climate-change-earth-too-hot-for-humans.html.

Wardle, Claire, and Hossein Derakhshan. "Thinking About 'Information Disorder': Formats of Misinformation, Disinformation, and Mal-information." In *Journalism, "Fake News" & Disinformation: Handbook for Journalism Education and Training*, ed. Cherilyn Ireton and Julie Posett, 43–54. Paris: UNESCO, 2018.

Watts, Jonathan. "Climatologist Michael E. Mann: 'Good People Fall Victim to Doomism. I Do Too Sometimes.'" *The Guardian*, February 27, 2021. https://www.theguardian.com/environment/2021/feb/27/climatologist-michael-e-mann-doomism-climate-crisis-interview.

Watts, Jonathan, Ashley Kirk, Niamh McIntyre, Pablo Gutiérrez, and Niko Kommenda. "Half World's Fossil Fuel Assets Could Become Worthless by 2036 in Net Zero Transition." *The Guardian*, November 4, 2021. https://www.theguardian.com/environment/ng-interactive/2021/nov/04/fossil-fuel-assets-worthless-2036-net-zero-transition.

Watts, Vanessa. "Indigenous Place–Thought and Agency Amongst Humans and Non Humans." *Decolonization: Indigeneity, Education & Society* 2, no. 1 (2013): 20–34. https://jps.library.utoronto.ca/index.php/des/article/view/19145.

Weise, Karen. "Amazon Settles with Activist Workers Who Say They Were Fired." *New York Times*, September 29, 2021. https://www.nytimes.com/2021/09/29/technology/amazon-fired-workers-settlement.html.

Welbourne, Dustin, and Will J. Grant. "What Makes a Popular Science Video on YouTube." *The Conversation*, February 24, 2015. http://theconversation.com/what-makes-a-popular-science-videoon-youtube-36657.

Wells-Barnett, Ida B. *Selected Works of Ida B. Wells-Barnett*. Ed. Henry Louis Gates Jr. London: Oxford University Press, 1991.

Wessler, Hartmut. *Habermas and the Media*. London: Wiley, 2019.

Westervelt, Amy. "How the Fossil Fuel Industry Got the Media to Think Climate Change Was Debatable." *Washington Post*, January 10, 2019. https://www.washingtonpost.com/outlook/2019/01/10/how-fossil-fuel-industry-got-media-thinkclimate- change-was-debatable/.

———. "We Can Tackle Climate Change If Big Oil Gets out of the Way." *The Guardian*, April 5, 2022. https://www.theguardian.com/environment/2022/apr/05/ipcc-report-scientists-climate-crisis-fossil-fuels.

Whitmarsh, Lorraine, and Adam Corner. "Tools for a New Climate Conversation: A Mixed-Methods Study of Language for Public Engagement Across the Political Spectrum." *Global Environmental Change* 42 (2017): 122–35. https://doi.org/10.1016/j.gloenvcha.2016.12.008.

Wilkins, Brett. "Green Groups' Petition Urges Social Media Platforms to Ban Big Oil Ads." Common Dreams, June 28, 2021. https://www.common

dreams.org/news/2021/06/28/green-groups-petition-urges-social-media-platforms-ban-big-oil-ads.

Wilson, Jason. "Social Media Disinformation on U.S. West Coast Blazes 'Spreading Faster Than Fire.'" *The Guardian*, September 14, 2020. https://www.theguardian.com/us-news/2020/sep/14/disinformation-oregon-wildfires-spreading-social-media.

Wong, Karen Li Xan, and Amy Shields Dobson. "We're Just Data: Exploring China's Social Credit System in Relation to Digital Platform Ratings Cultures in Westernised Democracies." *Global Media and China* 4, no. 2 (2019): 220–32. https://doi.org/10.1177/2059436419856090.

Wu, Tim. "Be Afraid of Economic Bigness. Be Very Afraid." *New York Times*, November 11, 2018. https://www.nytimes.com/2018/11/10/opinion/sunday/fascism-economy-monopoly.html?action=click&module=RelatedLinks&pgtype=Article.

———. *The Curse of Bigness: Antitrust in the New Gilded Age*. Columbia Global Reports. New York: Columbia University, 2018.

Yoder, Kate. "The Surprising Reasons Why People Ignore the Facts About Climate Change." *Grist*, July 28, 2020. https://grist.org/climate/the-surprising-reasons-why-people-ignore-the-facts-about-climate-change.

Zakrzewski, Cat. "Google Calls Itself Green. But It's Still Making Ad Money from Climate-Change Denial." *Washington Post*, December 16, 2021. https://www.washingtonpost.com/technology/2021/12/16/google-climate-change-denial-ads/.

Zelizer, Barbie, C. W. Anderson, and Pablo J. Boczkowski. *The Journalism Manifesto*. Cambridge: Polity, 2022.

Zeller, Tom, Jr. "Climate Talks Open with Calls for Urgent Action." *New York Times*, December 7, 2009. https://www.nytimes.com/2009/12/08/science/earth/08climate.html.

Zhou, Yanmengqian, and Lijiang Shen. "Confirmation Bias and the Persistence of Misinformation on Climate Change." *Communication Research* 49, no. 4 (2022): 500–523.

Zuboff, Shoshana. *The Age of Surveillance Capitalism: The Fight for a Human Future at the New Frontier of Power*. London: Profile, 2019.

———. "The Coup We Are Not Talking About." *New York Times*, January 29, 2020. https://www.nytimes.com/2021/01/29/opinion/sunday/facebook-surveillance-society-technology.html.

——. "You Are the Object of a Secret Extraction Operation." *New York Times*, November 12, 2021. https://www.nytimes.com/2021/11/12/opinion/facebook-privacy.html.

Zuckerman, Ethan. "The Case for Digital Public Infrastructure." Knight First Amendment Institute, Columbia University, 2020. https://knightcolumbia.org/content/the-case-for-digital-public-infrastructure.

——. "What Is Digital Public Infrastructure?" Center for Journalism & Liberty, November 17, 2020. https://www.journalismliberty.org/publications/what-is-digital-public-infrastructure#_ednref3.

INDEX

abolitionism, climate justice contrasted with, 157–59
accountability, of digital platforms, 147–48
activists, 110, 129; for climate justice, 112–28; data analytics utilized by, 111; harassment of, 120–21; incivility toward, 117–18, 121–22; journalism and, 56–57; legal action pursued by, 119; mainstream news outlets circumvented by, 124; media environment undermining, 108, 130–31; narrative influenced by, 121–22; social media utilized by, 124; within tech companies, 138; U.S. tracking, 78–79. *See also* Fridays for Future; No Dakota Access Pipeline
advertisers, 136–37, 201n6; big tech dominating, 82; communication influenced by, 146; disinformation turbocharged by, 99; journalism funded through, 150

agnotology, 39
AI. *See* artificial intelligence
algorithms, 19, 31–33, 65, 70, 80–87, 109, 111, 115, 124, 145, 147, 155
Algorithms of Oppression (Noble), 32
Amazon (e-commerce company), 13, 138
ambivalence, internet permeated by, 93–99
American Civil Liberties Union, 126
Apple (technology company), 13, 67, 68, 149
Araújo, Ernesto, 45
Arena, Christine, 136
Arendt, Hannah, 8, 72
Army Corps of Engineers, U.S., 123
Aronoff, Kate, 25–26
artificial intelligence (AI), 13, 31
Associated Press, 16, 54–55, 116–17
attention, information consuming, 82
audience, 14, 27–30, 98

Banks, Arron, 91
Barbaro, Michael, 32–36
Barnett, Julia, 120
Baudet, Thierry, 45
BBC (news outlet), 52–53, 150
bias, objectivity creating, 16–17, 46
Bidzíil, Shiyé, 125
big data, 13, 77, 80, 83–85, 101, 110
big tech: advertisers dominated by, 82; information infrastructures controlled by, 5; journalism and, 149, 155; Vice Media Group harmed by, 83
Black Lives Matter (movement), 29, 59–60, 105
Bolsonaro, Jair, 45
Bosworth, Andrew, 99
Bot Sentinel (tool), 89
BP (fossil-fuel company), 26, 43, 52, 136
Brown & Williamson (tobacco company), 40
Bullard, Robert, 4, 158–59

California Consumer Privacy Act, 145
Cambridge Analytica, 81, 144
Carbon Brief (website), 56, 57
"carbon footprint calculator," by BP, 26, 52
Carson, Rachel, 1–3, 167n1
Cassidy, Bill, 23
Center for Climate Change Communication, at George Mason University, 103

Chatham Asset Management (hedge fund), 149
CheckYourFact.com, 97
Chevron (fossil-fuel company), 43–45, 54, 136, 161
China, social media credit-score system in, 79
CleanCreatives (initiative), 137
Climateaction.tech, 138
climate change, 21, 55, 103; belief systems and, 28; ignorance and, 37; the individual blamed for, 26; media covering, 50–51; Trump dismissing, 39. *See also* Intergovernmental Panel on Climate Change
Climate Creatives (nonprofit), 99
climate crisis, 1–4, 14; framing of, 24–26; Inflation Reduction Act addressing, 161–62; information crisis intersecting with, 5–7, 12, 30, 102, 139, 162; journalism covering, 10, 17–18, 41–42, 49; mainstream news outlets reapproaching, 50; media landscape influencing, 10–11; racial disparity demonstrated by, 104–5; representations of, 60–61; scientists illuminating, 179n32. *See also* communication, on climate crisis
Climate Crisis Townhall, CNN, 89
climate justice, 28–29; abolitionism contrasted with, 157–59; activists for, 112–28; indifference reduced through, 129; journalism as

INDEX 247

focus of, 18; racial justice contrasted with, 158–59; Thompson Reuters Foundation focusing on, 59
Climate Science Information Center, by Facebook, 95–98
Climate Summits, by UN, 26–27, 43
CNN (media company), Climate Crisis Townhall by, 89
CO_2 Coalition, 97–98
Collapse of Western Civilization, The (Oreskes and Conway), 133–35
Commerce Committee, U.S. Senate, 137
Committee on Environment and Public Works, U.S. Senate, 23
Committee on Oversight and Reform, U.S. House, 135–36
communication, on climate crisis, 14, 103; advertisers influencing, 146; M. Boykoff on, 28; incivility impacting, 90–92; IPCC on, 24; media environment influencing, 36–37; noise impacting, 85–90
Communication Decency Act (1996), 147
Communications Act (1934), 142
Conference of the Parties (COP), by UN, 6, 43, 55
Conscious Advertising Network, 99
conspiracy theories, Twitter originating, 88
Consumer Privacy Act, California, 145
COP. *See* Conference of the Parties

Copenhagen, UN Climate Change Conference, 55
corporations: disinformation spread by, 28–29; the public not prioritized by, 135–36; publics disempowered by, 141; right-wing, 87–88. *See also* specific corporations
Costa, Maren, 138
Couldry, Nick, 83–84
counterpublics, water protectors exemplifying, 8
Covering Climate Now (journalism partnership), 49, 54
Cunningham, Emily, 138

Dakota Access Pipeline, at the Standing Rock Indian Reservation, 78, 90
data: big, 77, 83–85, 101; Cambridge Analytica harvesting, 81, 144; personal, 77–78, 144–46; stories contrasted with, 23
data analytics, activists utilizing, 111
data justice, 108–12
DDT (pesticide), 2
debate, appearance of, 47, 48
"Declaration of the Independence of Cyberspace, A" (Barlow), 34
Delaware River Basin, fracking banned in, 161
democracy: eco-authoritarianism defying, 195n12; platforms undermining, 65; profits prioritized over, 143

denialism, 51, 61–62, 169n23; Facebook propagating, 96–97; fossil-fuel companies spreading, 201n6; news outlets rewarding, 17; newsrooms struggling with, 52–53
DeSmog (news outlet), 56, 127
Dewey, John, 106
Dewey, Myron, 126
digital extraction, EU opposing, 145
digital infrastructures, publics shaped by, 101
Digital Smoke Signals (Facebook page), 126
disinformation, 100–101, 136; advertisers turbocharging, 99; corporations spreading, 28–29; ExxonMobil propagating, 42–43; Facebook expanding, 88; fossil-fuel companies spreading, 43, 73; journalism and, 42–50; noise created by, 89; right-wing corporations originating, 87–88; social media spreading, 80–81
Doctorow, Cory, 103
Dorsey, Jack, 32–33, 35–36
doubt, 40, 45–46, 64
Drone2bwild (Facebook page), 125
drones, 124–26
Dubuc, Nancy, 83

echo chambers, the internet creating, 81–82
eco-authoritarianism, democracy defied by, 195n12
Edelman (firm), 136

Eilish, Billie, 114
End of Nature (McKibben), 1
"End of Theory, The" (Anderson, Chris), 84–85
Energy Transfer Partners (company), 126–27
Engine No. 1 (hedge fund), 161
environmental justice: Indigenous people fighting for, 8, 123; inequality considered by, 26–27; TV news covering, 73. *See also* climate justice
Environmental Protection Agency, 2
epistemic rift, 76; institutions undermined by, 62–63; populists blamed for, 100–101; publics characterized by, 69–70
Eshelman, Robert, 53
European Union (EU), 145, 204n29
European Union Digital Services Act (1996), 147
Extinction Rebellion (group), 49–50
ExxonMobil (fossil-fuel company), 42–43, 52, 98, 136, 161

Facebook (social media), 67, 75, 119, 125–26, 155, 204n29; Climate Science Information Center by, 95–98; CO_2 Coalition proliferating through, 98; denialism propagating through, 96–97; disinformation expanding through, 88; Haugen exposing, 137–38; microtargeting by, 81;

presidential election influenced by, 81, 94, 144; as technology, 31
"Facebook Files, The" (reports), 137
Fairness Doctrine (1949), 142
FBI, 350.org discredited by, 79
FBI Portland, 88
Federal Aviation Administration, U.S., 125
FFF. *See* Fridays for Future
Fonda, Jane, 114
fossil-fuel companies, 80–81, 130; denialism spread by, 201n6; disinformation spread by, 43, 73; House of Representatives investigating, 135–36; the individual blamed by, 51–52; scientists competing with, 134–35; social media funded by, 99; universities divesting from, 160–61. *See also specific fossil-fuel companies*
fracking, Delaware River Basin banning, 161
freedom, rethinking of, 156–62
Fridays for Future (FFF), 38, 107, 110–13, 117–22; communication strategies of, 114; media-relations training by, 115–16; No DAPL contrasted with, 129–30

GDPR. *See* General Data Protection Regulation
Gebru, Timnit, 138
General Data Protection Regulation (GDPR), EU, 145, 204n29

Geofeedia (company), 126
geolocation tagging, police use of, 126
Ghosh, Amitav, 139
Giago, Tim, 124
global warming. *See* climate change
Goering, Laurie, 59
Goodman, Amy, 127
Google, 13, 31, 32, 67, 68, 69, 76, 91, 99, 138, 146, 149–50, 204n29
Greenpeace, 99
greentrolling, noise countered through, 118
Guardian, The (newspaper), 58, 79

Habermas, Jürgen, 71
Hague District Court, 161
Hansen, James, 47
harassment, 118; of activists, 120–21; protesters facing, 127; of scientists, 90–91
Haugen, Frances, 137–38
Hayes, Chris, 158
Hayhoe, Katharine, 28, 91
Hertsgaard, Mark, 10, 49
Hickman, Leo, 56, 57
Hip Hop Caucus, 99
Hope, Mat, 56
House of Representatives, U.S., 67, 70, 135–36
Houston Public Media, 44–45
How Climate Change Comes to Matter (Callison), 21
Hutchins Commission (1943), 142
hybrid media system, 12

ideologies, infrastructures embodying, 30–31
ignorance: climate change and, 37; doubt creating, 40; journalism influencing, 41, 50, 63–64; social media spreading, 64; tobacco industry producing, 39–40. *See also* disinformation
Imagined Communities (Anderson), 9
incivility: communication impacted by, 90–92; publics compromised by, 92; toward activists, 117–18, 121–22. *See also* harassment
Indigenous people, environmental justice fought for by, 8, 123
individual, the: climate change blamed on, 26; fossil-fuel companies blaming, 51–52; liberalism privileging, 157; libertarianism prioritizing, 159–60
Inflation Reduction Act, climate crisis addressed in, 161–62
InfluenceMap (nonprofit), 80, 96–97
InfoAmazonia (project), 111
information, attention consumed by, 82
information crisis, climate crisis intersecting with, 5–7, 12, 30, 102, 139, 162
information-deficit model, 21–22, 37, 48
information infrastructures, 43; big tech controlling, 5; liberalism influencing, 33–34; profits required by, 44; publics connected by, 5–6. *See also* big tech; the internet
infrastructures, ideologies embodied by, 30–31
Inglewood Oil Field, 161
Inhofe, James, 23
Inside Climate News (nonprofit), 55, 57, 59–60
Instagram (social networking service), 76, 89, 115, 120, 124, 127
Intercept, The (nonprofit), 126
Intergovernmental Panel on Climate Change (IPCC), UN, 4, 43, 96; on communication, 24; journalism following, 52; social science ignored by, 20
International Consortium of Investigative Journalism, 54
the internet: ambivalence permeating, 93–99; echo chambers created by, 81–82; Mexican Zapatistas utilizing, 11–12; pollution created by, 13; publics created through, 73–74; regulation reduced by, 142
Internet Research Agency (company), 128
IPCC. *See* Intergovernmental Panel on Climate Change

journalism, 66; activists and, 56–57; advertisers funding, 150; big tech and, 149, 155; changes in, 42; climate crisis covered by, 10,

17–18, 41–42, 49; climate justice as focus of, 18; collaboration within, 53–55, 154; debate upheld by, 47; disinformation and, 42–50; high-modern period of, 2–3; hybrid, 153–54; ignorance influenced by, 41, 50, 63–64; IPCC followed by, 52; networked technology impacting, 64–65, 74–75; objectivity jettisoned by, 53, 60–61; platforms overshadowing, 12; professional practices rethought by, 15–19, 51, 55–56, 68–69; profits negotiated by, 142; publics created through, 9; racial inequity illuminated by, 59–60; reforming of, 151–56; representations distorted by, 151; scientists and, 15–16, 55–56; sources of authority expanded by, 152–53; tech companies supporting, 148; user-generated content contrasted with, 86; violence against, 90–91

Judiciary Committee, of U.S. House of Representatives, 67, 70

justice: data, 108–12; media, 108–12, 131, 140; racial, 59, 158; social, 106–7. *See also* climate justice; environmental justice

Keystone XL (oil pipeline), 122
Khanna, Ro, 135–36
Koch Brothers, 97

Lambert, Grace, 116, 120–21
language, politics influenced by, 58
Larsson, Milane, 27
Law Enforcement Today (Facebook group), 88
Let's Get in the News (350.org), 115–16
liberalism, 140; the individual privileged in, 157; information infrastructures influenced by, 33–34; press freedom impacted by, 36
libertarianism, the individual prioritized by, 159–60
Line 3 (pipeline), 128
Lippmann, Walter, 22, 106
listening, struggling contrasted with, 28–30
live video, No DAPL enabled by, 124–25
Loeb, Vernon, 57, 59–60

mainstream news outlets: activists circumventing, 124; Carson writing for, 167n1; climate crisis reapproached by, 50
Mann, Michael, 26
Margolin, Jamie, 119–21
market fundamentalism, positivism competing with, 133–34
Marres, Noortje, 101, 106, 130
mass media: networked media transitioned to by, 41, 74; professional norms in, 16–17; public sphere dominated by, 72

material power, media demonstrating, 109
McClatchy (publishing company), 149–50
McKenna, Catherine, 91
McKibben, Bill, 1, 53, 160–61
media, 14; climate change covered by, 50–51; hybrid system of, 12; legacy news, 19, 141; material power demonstrated by, 109; publics shaped by, 15. *See also* mass media; news outlets; social media
Media and Climate Change Observatory, University of Colorado, 50–51
MediaClimate (research team), 6
media companies, the public disadvantaged by, 143
media environment, 5–6, 139–40, 154–55; activists undermined by, 108, 130–31; communication influenced by, 36–37; social justice influencing, 106–7
media justice, 108–12, 131, 140
media landscape, 5, 30, 141; climate crisis influenced by, 10–11; publics cultivated by, 7; Thunberg and, 57
Media Manifesto, The (Fenton), 141
Media Matters (non profit), 73
media strategies, of No DAPL, 112
Mexican Zapatistas (rebellion movement), 11–12
microtargeting, by Facebook, 80–81
misinformation. *See* disinformation

Moya-Smith, Simon, 124
Murdoch, Rupert, 68
myth of big data, 83–85, 101

Nakate, Vanessa, 116–17
the narrative: activists influencing, 121–22; facts framed by, 24; No DAPL shaping, 128
Nation, The (magazine), 54
National Rifle Association, 96
Native Sun News Today (news outlet), 124
"natural collectivity," social media feigning, 83–84
negative freedom, regulation discredited by, 34–35
networked media, mass media transitioning to, 41, 74
Networked Press Freedom (Ananny), 36
Networked Publics (Russell), 74
networked technology, impacts of, 64–65, 74–76
Neubauer, Luisa, 119
news deserts, Chevron exploiting, 44
news environment, transformation of, 41–42
news media, systemic inequities acknowledged by, 60
news outlets: denialism rewarded by, 17; Indigenous community-based, 124; Permian Proud mimicking, 44. *See also* mainstream news outlets
newsrooms, denialism struggled with in, 52–53

NGOs. *See* nongovernmental organizations
No Dakota Access Pipeline (No DAPL), 38, 107, 110–11; FFF contrasted with, 129–30; live video enabling, 124–25; media strategies of, 112, 123; the narrative shaped by, 128; police opposing, 124–27; Tar Sands Pipeline opposed by, 122
noise: communication impacted by, 85–90; disinformation creating, 89; greentrolling countering, 118; publics discerning through, 162
nongovernmental organizations (NGOs), 18–19

objectivity: bias created through, 16–17, 46; detachment caused by, 21–22; journalism jettisoning, 53, 60–61; scientists constrained by, 19–20
Ocasio-Cortez, Alexandria, 108, 118
Oceti Sakowin Camp, No DAPL, 123
oligopoly, of tech companies, 145–46

Panama Papers, 54
Pandora Papers, 54
Paris UN Climate Summit, 26–27, 43
Patagonia (brand), 114

Pelosi, Nancy, 108
Permian Proud (website), 44
personal data, surveillance capitalism and, 77–78
Phantom Public, The (Lippmann), 106
platforms, digital, 64–65; accountability of, 147–48; democracy undermined by, 65; journalism overshadowed by, 12; publishers contrasted with, 183n69. *See also* Facebook
police, No DAPL opposed by, 124–27
policing, predictive, 79
politics, language influencing, 58
pollution, the internet creating, 13
Pope, Kyle, 10
populists, epistemic rift blamed on, 100–101
positivism, market fundamentalism competing with, 133–34
Prakash, Varshini, 109
presidential election (2016), Facebook influencing, 81, 94, 155
press freedom, liberalism impacting, 36
profits: democracy not prioritized over, 143; information infrastructures requiring, 44; journalism negotiating, 142; the public not prioritized over, 9, 12, 29, 44, 70, 141–42
Program on Climate Change Communication, at Yale, 103

public, the: corporations not prioritizing, 135–36; media companies disadvantaging, 143; profits prioritized over, 9, 12, 29, 44, 70, 141–42; science distrusted by, 63

publics: automated, 76–77; corporations disempowered by, 141; counter, 8; digital infrastructures shaped by, 101; epistemic rift characterizing, 69–70; incivility compromising, 92; information infrastructures connecting, 5–6; the internet created by, 73–74; journalism creating, 9; media landscape cultivating, 7; media shaping, 15; networked, 73–74, 93; networked technology impacting, 76; noise discerned by, 162; sociomateriality shifting, 71–83

public sphere, mass media dominating, 72

publishers, platforms contrasted with, 183n69

racial justice, 59, 158–59

Reckoning (Callison and Young), 151–52

redlining, technological, 79

regulation, 204n29; the internet reducing, 142; negative freedom discrediting, 34–35; surveillance economics targeted by, 144–45

representations: of climate crisis, 60–61; communication researchers prioritizing, 22–23; journalism distorting, 151

Rich, Nathaniel, 25

right-wing corporations, disinformation originating from, 87–88

right-wing leaders, science dismissed by, 45–46

Ritz, Gerry, 91

Rossiter, Caleb, 98

Ruffalo, Mark, 114

science, 2, 95–98; *The Guardian* prioritizing, 58; professional practices of, 19–22; the public distrusting, 63; right-wing political leaders dismissing, 45–46; social, 20

scientist citizens, 56

scientists: climate crisis illuminated by, 179n32; fossil-fuel companies competing with, 134–35; harassment of, 90–91; journalism and, 15–16, 55–56; objectivity constraining, 19–20

Seattle Independent Media Center, 12

Silent Spring (Carson), 1–3, 167n1

Sinclair, Upton, 3

Singer Associates (public relations firm), 44

Snowden, Edward, 144

social justice, media environment influenced by, 106–7

social media: activists utilizing, 124; disinformation spread on, 80–81; fossil-fuel companies funding, 99; ignorance spread through, 64; "natural collectivity" feigned on, 83–84. *See also* Facebook
social media credit-score system, in China, 79
"Standards for Media" (Extinction Rebellion), 49–50
Standing Rock Indian Reservation, 78, 90, 126
Standing Rock Sioux, Army Corp of Engineers sued by, 123
stories, data contrasted with, 23
Structural Transformation of the Public Sphere, The (Habermas), 71
struggling, listening contrasted with, 28–30
Sunrise Movement, 108, 109
Surveillance Capitalism (Zuboff), 144–45
surveillance capitalism, personal data and, 77–78
surveillance economics, regulation targeting, 144–45
Swedish Society for Nature Conservation (nonprofit), 78
systemic inequities, news media acknowledging, 60

Tar Sands Pipeline, No DAPL opposing, 122
TC Energy, Keystone XL terminated by, 122
tech companies, 87, 94; activists within, 138; ambivalence of, 95; antitrust concerns about, 67–68; big data benefiting, 85; House Judiciary Committee investigating, 67, 70; journalism supported by, 148; oligopoly of, 145–46; structural solutions avoided by, 36. *See also* big tech
Tech Workers Coalition, 138
Telecommunications Act (1996), 142
Thompson Reuters Foundation, 59
350.org (organization), 1, 79, 115–16
Thunberg, Greta, 57, 91, 112–13, 116–18
TigerSwan (firm), 126
TikTok, 76, 99, 115
tobacco industry, 39–40, 67
transparency, Dorsey prioritizing, 33, 35–36
Trump, Donald, 35, 39, 46, 117–18
TV news, environmental justice covered by, 73
Twitter, 32, 35, 67, 75, 88–89, 120
tyranny of the loudest, 31–32

UN. *See* United Nations
"Uninhabitable Earth, The" (Wallace-Wells), 24–25
United Nations (UN): Climate Change Conference by, 55; Climate Summits by, 26–27, 43; COP by, 6; Paris Climate Summit by, 26–27. *See also* Intergovernmental Panel on Climate Change

United States (U.S.): activists tracked by, 78–79; Army Corp of Engineers for, 123; Federal Aviation Administration of, 125; House of Representatives of, 67, 70, 135–36; Senate of, 23, 137; Supreme Court of, 128

universities, fossil-fuel companies divested from by, 160–61

"vibrant materiality," J. Bennett on, 156–57

Vice, 53, 125–26

Vice Media Group, 83

Viner, Katharine, 58

"Violence of the Market, The" (Pickard), 148

water protectors, 8, 125–26

We Are the Weather (Foer), 25

Wells, Ida B., 3

WhatsApp, 114, 120

Wired (magazine), 84, 118–19

Woods, Darren, 136

You Are Here (Phillips and Milner), 29–30

YouTube, 32, 86–87, 115, 146

Zapatistas, Mexican (rebellion movement), 11–12

Zuboff, Shoshana, 77, 144–45, 147

Zuckerberg, Mark, 31, 34, 94–95, 119, 170n34

GPSR Authorized Representative: Easy Access System Europe, Mustamäe tee
50, 10621 Tallinn, Estonia, gpsr.requests@easproject.com

www.ingramcontent.com/pod-product-compliance
Lightning Source LLC
Chambersburg PA
CBHW022044290426
44109CB00014B/978